万达商业规划 2016

WANDA COMMERCIAL PLANNING

万达商业规划研究院
万达商业地产设计中心　　主编
万达商业地产技术研发部

中国建筑工业出版社

EDITORIAL BOARD MEMBERS
编委会成员

主编单位
万达商业规划研究院
万达商业地产设计中心
万达商业地产技术研发部

规划总指导
王健林

执行编委
赖建燕 曲晓东 于修阳 黄国斌 王顺成 叶宇峰 尹强
朱其玮 林树郁 门瑞冰 方伟 侯卫华 张东光 杨旭
沈文忠 李浩 季元 王群华 毛晓虎 石亮 陈文娜
刘大伟 朱镇北 秦鹏华

王福魁 孙培宇 曾静 风雪昆 王艳明

参编人员
范珑 陈海亮 曹春 阎红伟 黄引达 薛勇 刘征 文善平
刘江 王治天 薛瑜 刘安 王权 荣万斗 陆峰 赵青扬
王宇石 刘佩 高振江 李斌 曹国峰 昌燕 叶啸 周昳晗
武春雨 胡延峰 张鹏翔 董华维 赵龙 黄建好 边界
王文广 朱欢 屈娜 刘保亮 孙佳宁 李彬 张堃 石路也
董明海 路清淇 苏仲洋 范群立 方芳 蒲峰

马红 兰勇 章宇峰 罗沁 王玉龙 张振宇 黄路 黄勇
徐立军 王朝忠 郭扬 李小强 黄涛 葛宁 冯俊 赵昂
阳舒华 孙志超 俞小华 安云泽 周升森 张悦 马长宁
张宁

李晅荣 车心达 林彬 赵宁宁 谭喜峰 张雪晖 杨磊
马刚 栾赫 潘亮 覃涛 彭亚飞 刘晓敏 李春阳 洪剑
漆国强 马申申 孙一琳 李鹏 刘敏 纪文青 顾东方
李子强 庞庆 王凡 李江涛 冯董 龚芳 宋波 霍雪影
邵强 张克 李万顺 梁国涛 刘晓波 张飚 李捷 高建航
袁喆 王清文 王书研 姚建刚 孙海龙 王奕 杨春龙
沈余 宋雷 吕鲲 桑伟 张晓冬 王吉 晁志鹏 余斌
孟晗 王进纯 徐小莉 主佳 何志勇 刘向阳 罗琼 邹洪
戴欣欣 方文奇 李华 王凯 孟祥宾 栾海 王静 段堃

范志满 杨健珊 梅帆 桑国安 杜文天 王雅斯 陈晖
任睿 高维 王嵘 周鹏 张涛 朱迪 张洋 熊厚 杜鹏展
张顺 张黎明 陆轩 杨琳 冯晓芳 高霞 李民伟 潘鸿岭
戚士林 赵洪斌 都晖 林涛 党恩 蓝毅 张琳 张鹤

校对
张东光 胡延峰 叶啸 薛瑜 万志斌 刘敏 张晓冬
王书研 杨春龙 王蓉菲 张堃 王雪娇 张蕾 马丽
郭家桢 袁文卿

英文校对
余刚 方芳 朱欢 高建航 袁喆 王书研 沈洋 郭家桢

CHIEF EDITORIAL UNITS
Wanda Commercial Planning & Research Institute
Wanda Commercial Estate Design Center
Wanda Commercial Building Intelligence Department

GENERAL PLANNING DIRECTOR
Wang Jianlin

EXECUTIVE EDITORIAL BOARD MEMBERS
Lai Jianyan, Qu Xiaodong, Yu Xiuyang, Huang Guobin, Wang Shuncheng, Ye Yufeng, Yin Qiang, Zhu Qiwei, Lin Shuyu, Men Ruibing, Fang Wei, Hou Weihua, Zhang Dongguang, Yang Xu, Shen Wenzhong, Li Hao, Ji Yuan, Wang Qunhua, Mao Xiaohu, Shi Liang, Chen Wenna, Liu Dawei, Zhu Zhenbei, Qin Penghua

Wang Fukui, Sun Peiyu, Zeng Jing, Feng Xuekun, Wang Yanming

PARTICIPANTS
Fan Long, Chen Hailiang, Cao Chun, Yan Hongwei, Huang Yinda, Xue Yong, Liu Zheng, Wen Shanping, Liu Jiang, Wang Zhitian, Xue Yu, Liu An, Wang Quan, Rong Wandou, Lu Feng, Zhao Qingyang, Wang Yushi, Liu Pei, Gao Zhenjiang, Li Bin, Cao Guofeng, Chang Yan, Ye Xiao, Zhou Yihan, Wu Chunyu, Hu Yanfeng, Zhang Pengxiang, Dong Huawei, Zhao Long, Huang Jianhao, Bian Jie, Wang Wenguang, Zhu Huan, Qu Na, Liu Baoliang, Sun Jianing, Li Bin, Zhang Kun, Shi Luye, Dong Minghai, Lu Qingqi, Su Zhongyang, Fan Qunli, Fang Fang, Pu Feng

Ma Hong, Lan Yong, Zhang Yufeng, Luo Qin, Wang Yulong, Zhang Zhenyu, Huang Lu, Huang Yong, Xu Lijun, Wang Chaozhong, Guo Yang, Li Xiaoqiang, Huang Tao, Ge Ning, Feng Jun, Zhao Ang, Yang Shuhua, Sun Zhichao, Yu Xiaohua, An Yunze, Zhou Shengsen, Zhang Yue, Ma Changning, Zhang Ning

Li Xuanrong, Che Xinda, Lin Bin, Zhao Ningning, Tan Xifeng, Zhang Xuehui, Yang Lei, Ma Gang, Luan He, Pan Liang, Qin Tao, Peng Yafei, Liu Xiaomin, Li Chunyang, Hong Jian, Qi Guoqiang, Ma Shenshen, Sun Yilin, Li Peng, Liu Min, Ji Wenqing, Gu Dongfang, Li Ziqiang, Pang Qing, Wang Fan, Li Jiangtao, Feng Dong, Gong Fang, Song Bo, Huo Xueying, Shao Qiang, Zhang Ke, Li Wanshun, Liang Guotao, Liu Xiaobo, Zhang Biao, Li Jie, Gao Jianhang, Yuan Zhe, Wang Qingwen, Wang Shuyan, Yao Jiangang, Sun Hailong, Wang Yi, Yang Chunlong, Shen Yu, Song Lei, Lv Kun, Sang Wei, Zhang Xiaodong, Wang Ji, Chao Zhipeng, Yu Bin, Meng Han, Wang Jinchun, Xu Xiaoli, Zhu Jia, He Zhiyong, Liu Xiangyang, Luo Qiong, Zou Hong, Dai Xinxin, Fang Wenqi, Li Hua, Wang Kai, Meng Xiangbin, Luan Hai, Wang Jing, Duan Kun

Fan Zhiman, Yang Jianshan, Mei Fan, Sang Guoan, Du Wentian, Wang Yasi, Chen Hui, Ren Rui, Gao Wei, Wang Rong, Zhou Peng, Zhang Tao, Zhu Di, Zhang Yang, Xiong Hou, Du Pengzhan, Zhang Shun, Zhang Liming, Lu Xuan, Yang Lin, Feng Xiaofang, Gao Xia, Li Minwei, Pan Hongling, Qi Shilin, Zhao Hongbin, Du Hui, Lin Tao, Dang En, Lan Yi, Zhang Lin, Zhang He

PROOFREADERS
Zhang Dongguang, Hu Yanfeng, Ye Xiao, Xue Yu, Wan Zhibin, Liu Min, Zhang Xiaodong, Wang Shuyan, Yang Chunlong, Wang Rongfei, Zhang Kun, Wang Xuejiao, Zhang Lei, Ma Li, Guo Jiazhen, Yuan Wenqing

ENGLISH VERSION PROOFREADERS
Yu Gang, Fang Fang, Zhu Huan, Gao Jianhang, Yuan Zhe, Wang Shuyan, Shen Yang, Guo Jiazhen

CONTENTS
目录

PREFACE 010
序言

BUSINESS TRANSFORMATION BASICALLY COMPLETED 012
基本实现企业转型

WANDA COMMERCIAL PLANNING 2016 014
万达商业规划 2016

WANDA CITY 018
万达城

WANDA CITY NANCHANG
南昌万达城

01 MASTER PLAN OF WANDA CITY NANCHANG 020
南昌万达城总体规划

02 PUBLIC LANDSCAPE PLANNING OF WANDA CITY NANCHANG 024
南昌万达城公共景观规划

03 WANDA MALL NANCHANG 026
南昌万达茂

04 HOTEL COMPLEX OF WANDA CITY NANCHANG 036
南昌万达城酒店群

05 WANDA CITY NANCHANG - BAR STREET 056
南昌万达城酒吧街

06 LAKESIDE GREEN BELT OF THE JIULONG LAKE, WANDA CITY NANCHANG 059
万达南昌九龙湖滨湖绿带

WANDA CITY HEFEI
合肥万达城

01 MASTER PLAN OF WANDA CITY HEFEI 060
合肥万达城总体规划

02 WANDA MALL HEFEI 064
合肥万达茂

03 HOTEL COMPLEX OF WANDA CITY HEFEI 074
合肥万达城酒店群

04 BAR STREET OF HOTEL COMPLEX IN WANDA CITY HEFEI
合肥万达城酒店群酒吧街 096

WANDA PLAZA
万达广场 098

01 BEIJING FENGTAI HIGH-TECH PARK WANDA PLAZA 100
北京丰台科技园万达广场

02 BEIJING HUAIFANG WANDA PLAZA 110
北京槐房万达广场

03 CHENGDU SHUDU WANDA PLAZA 118
成都蜀都万达广场

04 JI'NAN HIGH-TECH WANDA PLAZA 126
济南高新万达广场

05 YANTAI DEVELOPMENT ZONE WANDA PLAZA 134
烟台开发区万达广场

06 CHAOYANG WANDA PLAZA 142
朝阳万达广场

07 BOZHOU WANDA PLAZA 150
亳州万达广场

08 JIXI WANDA PLAZA 156
鸡西万达广场

09 MUDANJIANG WANDA PLAZA 162
牡丹江万达广场

10 URUMQI JINGKAI WANDA PLAZA 168
乌鲁木齐经开万达广场

11 DEYANG WANDA PLAZA 172
德阳万达广场

D WANDA HOTEL 176
万达酒店

01 WANDA REIGN SHANGHAI 178
上海万达瑞华酒店

02 WANDA VISTA ZHENGZHOU 184
郑州万达文华酒店

03 WANDA VISTA URUMQI 188
乌鲁木齐万达文华酒店

04 WANDA REALM SIPING 192
四平万达嘉华酒店

05 WANDA REALM XINING 196
西宁万达嘉华酒店

06 WANDA REALM BOZHOU 200
亳州万达嘉华酒店

07 WANDA REALM YIWU 204
义乌万达嘉华酒店

08 WANDA REALM SHANGRAO 208
上饶万达嘉华酒店

E PROPERTIES FOR SALE 212
销售类物业

01 EXHIBITION CENTER OF WANDA CITY CHONGQING 214
重庆万达城展示中心

02 EXHIBITION CENTER OF WANDA NO.1 CHENGDU 224
成都万达一号展示中心

03 HEPING EXHIBITION CENTER OF WANDA CITY GUILIN 232
桂林万达和平展示中心

04 DEMONSTRATION AREA OF YONGCHUAN WANDA PALACE 242
永川万达华府示范区

05 PROTOTYPE SHOP DEMONSTRATION AREA OF WANDA CITY CHENGDU - GREEN TOWN & WATER STREET 250
成都万达城商铺样板示范区——青城水街

06 COMMERCE & RESIDENCE BLOCK OF ZHANGZHOU TAISHANG WANDA PLAZA 256
漳州台商万达广场商墅

F INTERNATIONAL SHOPPING PLAZA CONCEPT COMPETITION
"概念商业广场"国际设计竞赛 — 260

G DESIGN & CONTROL
设计及管控 — 276

R&D EVENTS OF WANDA BIM GENERAL CONTRACTING MANAGEMENT — 278
"万达 BIM 总发包管理模式"研发大事记

SUMMARY FOR R&D RESULTS OF WANDA BIM GENERAL CONTRACTING MANAGEMENT IN 2016 — 280
2016"万达 BIM 总发包管理模式"研发成果总结

SUMMARY FOR IMPLEMENTATION OF DESIGN GENERAL CONTRACTING MODEL — 284
"设计总发包模式"推行小结

R&D AND APPLICATION OF HUIYUN SYSTEM V3.0 — 287
"慧云系统"（3.0 版）的研发及应用

R&D AND OPERATION & MAINTENANCE MANAGEMENT OF HUIYUN SYSTEM V3.0 — 289
"慧云系统"（3.0 版）的研发及运维管理

BUILD A WORLD-CLASS CULTURAL TOURIST DESTINATION BASED ON REGIONAL CULTURE - INNOVATIVE PRACTICE OF WANDA CULTURAL TOURISM CITY AND WANDA MALL — 292
依托地域文化，打造世界级文化旅游目的地
——万达文化旅游城与万达茂的创新实践

WANDA IN THE EYES OF ALL CIRCLES — 294
各界看万达

H PROJECT INDEX
项目索引 — 296

序言 PREFACE

BUSINESS TRANSFORMATION BASICALLY COMPLETED
基本实现企业转型

（一）服务业收入、净利润大于地产

2016年万达集团服务业收入占比55%，历史上首次超过地产。服务业净利润（未经审计）占比超过60%，也大于地产开发利润。万达提前一年实现了转型阶段目标。

（二）成功转型轻资产

2016年开业的50个万达广场，已有21个属于轻资产。2016年第4季度，万达商业与中信信托、民生信托、富力集团等签约90个万达广场、共1050亿元的投资合同。2017年至2019年，每年开业交付30个万达广场给投资方，净租金双方分成。2016年万达商业轻资产还探索出一种新模式，叫"合作类"万达广场，就是对方出地又出钱，万达负责设计、建设指导、招商运营，净租金双方7：3分成。

不是对方所有项目我们都同意做，要先对项目进行筛选，对设计、建造进行指导，再招商、运营。设计也好，建造监理也好，招商也好，费用都是对方出。

"合作类"万达广场模式可以说是万达轻资产模式的最高级形式，连资本化环节都省掉了，还解决了集体用地问题。这种合作模式不但能分得租金，省去资本化，还解决集体用地问题，是万达商业的重大创新。

我多年前曾讲过一句话，企业经营的最高境界就是"空手道"。但这个空手道可不是骗子，是有了品牌、有了能力，别人找上门来，你一分钱不出，凭品牌就能挣钱。比如说酒店管理公司，因为有品牌，你找他管理，设计你出钱、人员工资你付，不管酒店盈亏，管理公司都分钱。世界上还有很多"特许经营"都是接近"空手道"，要做到这个境界是极难的。2016年万达商业"合作类"万达广场项目实现重大突破，2015年只签了一个，2016年不但签了18个，还开业一个"合作类"项目——就是北京丰台区的槐房万达广场。它位置较好，开业效果也不错。今年"合作类"万达广场能开业5个，2018年以后能每年开业20个左右。

（三）新兴产业高速增长

去年地产收入下降25%，但是万达集团总的收入还实现增长，就是因为文化、网络这些新兴产业增速大大高于地产。万达新兴产业增长快、前景好，相当一部分产业在国内还具有唯一性，比如万达网络科技，就是中国唯一的"实体商业+互联网"模式企业。万达体育已打造好几个自有IP赛事。这些赛事在国内没有竞争对手。乐观估计，2018年文化集团就会成为万达又一个千亿级企业。其实，如果文化集团不剥离旅行社，今年就接近千亿规模了。

（四）租金大幅增长

万达商业收入下降，但净利润为增长，一个重要原因是租金增长3成，虽然不是唯一因素，就是因为租金净利润率高。

——摘自王健林董事长《万达集团2016年年度工作报告》

万达集团董事长　王健林

(I) SERVICES CONTRIBUTED MORE REVENUE AND PROFIT THAN REAL ESTATE

In 2016, services contributed 55% of the Group's total revenue, surpassing the real estate business for the first time, and the net profit (unaudited) from services in the year was also greater than that from the real estate business. Wanda has accomplished its preliminary goal of transformation one year in advance.

(II) SUCCEEDED IN ITS SHIFT TO THE ASSET-LIGHT STRATEGY

Of the 50 Wanda Plazas opened in 2016, 21 were asset-light projects. In the fourth quarter of 2016, Dalian Wanda Commercial Properties signed contracts with companies including China CITIC Bank, China Minsheng Trust and R&F Properties over approximately 90 Wanda Plazas with a combined value of RMB 105 billion. From 2017 to 2019, Wanda will deliver 30 Wanda Plazas to investors every year and share net rent revenues with them. Even more encouragingly, in 2016 we have developed a new asset-light format called "collaboration-category Wanda Plazas", where investors provide both land and funds and Wanda is responsible for design construction advice, merchant recruitment and operation, with the investor and Wanda sharing the net rent on a 70/30 basis.

However, we don't do all projects offered to us but are selective according to our own criteria before commencing the projects going through the entire process from design and construction guidance to merchant recruitment and operation. Be it design, construction supervision or merchant recruitment, all related costs are borne by investors.

The collaboration-category Wanda Plaza model is arguably the highest form of Wanda's asset-light model, where not only the capitalization part is omitted but the issue related to collectively owned land is satisfactorily solved.

Years ago, I made the observation that the highest form of business operation should be like the art of "empty hand fighting". However, this is not the same thing as catching a white wolf with bare hands; instead, it means that with a strong brand and unique expertise, you will be approached for cooperation and making money together, where your brand is your main asset. An example is a hotel management brand that manages your properties for a fee and makes money regardless of whether or not you make a profit from the properties. There are many other forms of the "empty hand art" in the world, but it is very difficult to establish one. In 2016, Dalian Wanda Commercial Properties made a major breakthrough in the development of collaboration-category Wanda Plazas. Compared to 2015 when it only signed one such project, in 2016 it not only signed 18 but put one into operation, i.e. the favourably located Hua fang Wanda Plaza in Fengtai District in Beijing which has performed strongly since opening. We will open five collaboration-category Wanda Plazas in 2017 and approximately 20 every year from 2018.

(III) STRONG GROWTH IN EMERGING INDUSTRIES

In spite of a 25% decrease in revenues from the real estate business, Wanda Group maintained a positive revenue growth, driven by strong growth in emerging industries such as culture and network. Wanda has enjoyed a strong momentum with encouraging outlook in the emerging industries, with quite a few operations that stand in a class of their own, such as Wanda Network Technology that is the only "real economy + Internet" enterprise in China. Wanda Sports has developed a few self-owned sports events that have no competitor in China. According to optimistic estimates, Wanda Cultural Industry Group will reach a scale of RMB 100 billion in 2018. In fact, if the travel agency business had not been stripped, Wanda Cultural Industry Group would have been close to the RMB 100 billion mark this year.

(IV) RENT REVENUE HAS INCREASED SUBSTANTIALLY

Regardless of a decrease in total revenue, Dalian Wanda Commercial Properties achieved a positive growth in net profit, driven prominently by a 30% increase in rent revenue, among other factors.

—— Excerpted from Chairman Wang Jianlin's *Work Report of Wanda Group in 2016*

Chairman of Wanda Group　Wang Jianlin

WANDA COMMERCIAL PLANNING 2016
万达商业规划 2016

| 万达商业地产高级副总裁　赖建燕

前言

2016年，是万达集团宣布企业转型，开始轻资产模式的第一年，也是万达商业规划院锐意进取、变革创新的一年。

一、标准化研发+BIM

自2008年以来，万达商业规划持续进行标准化研发与积累。到2016年，在标准化与BIM研发方面取得突破性进展；标准化与BIM总发包的应用成为今年最亮丽的特点。2016年万达广场的标准化率已达70%，部分标准化产品实现产业化。万达广场采光顶已实现从标准化到产业化的进程。2016年开业的50个万达广场100%采用标准化产品，每个项目节省工期超过25天。设计材料封样库首次采用业主、设计总包和施工总包共建、共享、共用的模式，减少了以往封样工作中大量繁重重复的差旅和对接，极大地提升了工作效率。标准化、产业化的实施，是万达广场开业数量逐年增长、开业品质稳步提升的重要保障，不仅带来巨大经济和社会效益，更与国家推行的产业化政策精神完全吻合（如图1所示）。

二、设计总包模式

2016年，万达集团率先在国内商业项目全面推行设计总包，文旅项目试点推行设计总包。实施设计总包后，每个万达广场项目对接的设计单位数量、多方会议和协调对接工作量均大幅下降，仅会议数量就从原来的每项目平均600次下降为不到100次。为确保设

FOREWORD

In 2016, Wanda Group announced its transformation and started the first year of its asset-light model. In the same year, Wanda Commercial Planning Institute forged ahead through reform and innovation.

I. STANDARDIZATION R&D + BIM

Since 2008, Wanda Commercial Planning has been continuously engaged in R&D and practicing of "standardization". By 2016, breakthroughs have been made in R&D of both standardization and BIM; the application of standardization and BIM Turnkey Contract has been the highlight of the year. In this year, the "standardization" rate of Wanda Plaza projects reached 70%, and part of standardized products had been industrialized. The day-lighting roof design for Wanda Plaza projects has realized the transformation from standardization to industrialization. All the 50 Wanda Plaza projects which opened in 2016 have adopted standardized products, thus saving construction period by over 25 days for each project. The design material sample tank initiates the "jointly established, shared and utilized" model by the Owner, Design General Contractor and General Contractor for construction, which reduces burdensome business travel times and docking workload related to sample sealing and greatly improves work efficiency. "Standardization" and "Industrialization" guarantee improvement of Wanda Plaza projects which have been put into operation in terms of both quantity and quality, bringing about huge economic and social benefits for the Group and standing in line with the industrialization policy proposed by the Chinese government (as shown in Fig. 1).

II. GENERAL DESIGN CONTRACTING MODEL

In 2016, Wanda Group takes the lead in comprehensively implementing the "general design contracting model" in Chinese commercial projects, and in trial application of the model in cultural tourism projects. After the model is implemented, the number of designers, multiple-party meetings and coordination as required by each Wanda Plaza project are significantly reduced, with meetings alone dropped from the original 600 times on average to less than 100 times. With the aim to ensure design quality,

	序号	完成时间	标准化事项	标准化成果	阶段标准化率	累计标准化率
专项模块	1	2012	建造标准（限额设计）	《万达广场定额设计技术标准》（2012版）	5%	5%
	2	2015-I	通用标准施工图集	《通用标准施工图集》	10%	15%
	3	2015-II	BIM标准单元模块	采光顶、制冷机房、慧云机房标准单元模块	5%	20%
	4	2015-III	专项标准模块化	立面、内装、景观等效果类标准方案	10%	30%
	5	2015-2016	效果类标准样板库	效果类各级建造标准设计封样样板库及电子库	10%	40%
BIM模块	6	2016-II	BIM标准构件库 BIM标准模型	BIM标准构件库、WD-BIM标准模型	30%	70%
平面模块	7	2016-I	设计模块组合平面	A-、B、B-版共10个标准组合平面		

图1 标准化进程

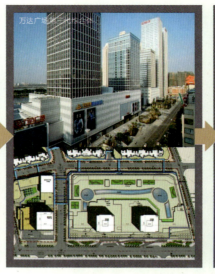

图2 四代产品演进图

计质量，还引入第三方进行审图把关，图纸问题数量也由年初的平均每个项目217个下降为116个。设计总包模式的推行，与国际项目管理模式接轨，与住房和城乡建设部将推行的"建筑师负责制"政策精神相吻合，大幅提升了工作效率，在国内建设行业的率先实践，对行业起到积极示范作用！万达的标准化与设计总包管控模式，共同奠定了BIM总发包管理模式的坚实基础。

三、万达茂（商业升级）

2016年，万达商业规划在第一代产品长白山国际度假区、西双版纳国际度假区的基础上，首次建成两个第二代产品——南昌万达茂和合肥万达茂。这两个"茂"首次将海洋乐园、电影乐园、宝贝王、电影院线、大玩家等主题娱乐业态纳入原有购物中心形态，从而创造了购物中心新形态，使其成为集购物、休闲、体验、美食、娱乐、文化、旅游为一体的中国首批文旅商综合体，是万达对国内商业建筑和商业广场的一次大胆探索。这一探索具有里程碑式的意义，从此拉开了万达茂大规模建设和开业的序幕：2017年万达计划开业南宁万达茂、哈尔滨万达茂，2018年计划开业南京万达茂、青岛万达茂，2019年计划开业上海万达茂、广州万达茂、无锡万达茂、昆明万达茂，2020年计划开业重庆万达茂、成都万达茂、济南万达茂（如图2所示）。

四、概念广场创新

为对商业规划进行总结和提升，应对商业社区小型化、电子商务对大型购物中心和零售业的冲击，2016年万达商业规划策划举行"第二届概念商业广场国际设计大赛"。大赛由中国建筑文化研究会与万达商业规划研究院联合主办，旨在通过搭建国际竞赛平台，吸引全球行业精英，探索在设计中将"商业广场空间、商户、消费者"三大元素有机结合，重塑"商业广场空间、业态、运营"之间的生态关系。大赛邀请国际著名建筑大师丹尼尔·李伯斯金、中国工程院院士、东南大学教授王建国等中外重量级行业嘉宾担任评委（如图3所示）。此次竞赛是万达商

"the third party" is employed to review drawings. As a result, the number of drawings with problems of each project has dropped from 217 at the beginning of 2016 to 116. The "general design contracting model" adopted by Wanda Group conforms to the "architect responsibility system" to be implemented by the Ministry of Housing and Urban-Rural Development of China, and is compatible with international project management pattern. In this sense, the model substantially improves work efficiency and plays as an exemplary role in the Chinese construction industry. Wanda's standardization and general design contracting model jointly contribute to a solid foundation for BIM Turnkey Contract Model.

III. WANDA MALL (COMMERCIAL UPGRADE)

In 2016, Wanda Commercial Planning, based on the first generation products (Changbai Mountain International Resort and Xishuangbanna International Resort), firstly completed its second generation products, which were Nanchang Wanda Mall and Hefei Wanda Mall. For the first time, these two Malls present a new form by adding theme entertainment business type into the original shopping mall, such as Ocean Park, Movie Park, Kids Place, Cinema and Super Player. Thus, both malls emerged as China's first cultural tourism commerce complexes that bring together shopping, leisure, experience, food, entertainment, culture and tourism, and mark a bold exploration of Wanda into domestic commercial buildings and commercial plazas. This exploration is a milestone that unveils the prelude of scale construction and opening for Wanda Malls. In 2017, Wanda plans to open Wanda Malls in Nanning and Harbin; in 2018, in Nanjing and Qingdao; in 2019, in Shanghai, Guangzhou, Wuxi and Kunming; in 2020, in Chongqing, Chengdu and Jinan (see Fig. 2).

IV. INNOVATION OF CONCEPT COMMERCIAL PLAZA

In order to summarize and upgrade commercial planning, respond to the impact of miniaturization of commercial communities and e-commerce on grand shopping malls and retailers, Wanda Commercial Planning planned to hold the "2nd International Shopping Plaza Concept Competition" in 2016. The competition is jointly organized by Architecture and Culture Society of China and Wanda Commercial Planning and Research Institute (WCPRI). With this international competition platform, it aims to gather the worldwide industry elites for exploring organic combination of three elements-shopping plaza space, merchants and consumers when designing and for reshaping ecological relationship among "shopping plaza space, business type, and operation". The competition is honored to have industry-renowned judges at home and abroad, such as Daniel Libeskind, a world renowned architect, and Wang Jianguo, Academician of Chinese Academy of Engineering and Professor of Southeast University (see Fig. 3). This is the 2nd international design competition organized by WCPRI following the successful holding of the first such competition in 2015. It is also an aggressive exploration

图3 "第一届概念商业广场国际建筑设计竞赛"部分嘉宾及评委合影

业规划研究院继2015年首次成功举办"概念商业广场"国际建筑设计竞赛后的又一次国际设计竞赛活动,也是万达商业规划每年在进行大量实体广场规划设计的同时,对未来广场发展方向、商业形态变化和时尚潮流进行的积极探索,体现了万达规划的社会责任。

made by Wanda Commercial Planning in terms of future development direction of plaza, business pattern change and fashion trend, while planning and designing a large number of off-line plazas each year. The competition well exhibits the social responsibility Wanda planning is willing to shoulder.

五、销售物业创新

万达销售物业最初是以纯开发项目形式出现,随着一代二代万达广场的出现,逐渐形成了万达独有的特点:住宅以高层住宅为主,商铺、公寓、写字楼并存。伴随文旅项目的兴起,万达销售物业形式日趋丰富,住宅形成别墅、洋房、高层完整系列。2016年,我们更进一步在销售物业中开发出合院、别院、叠院等新产品,进一步丰富和补充了原有产品类型,得到市场认可(如图4所示)。

V. INNOVATION OF PROPERTIES FOR SALE

With the experience of building the first and second generation Wanda Plazas, Wanda properties for sales gradually grow its unique characteristics from initially pure development projects: mainly high-rise residence coexisting with shops, apartments and office buildings. With the rise of cultural tourism projects, Wanda properties for sale witness increasingly enriched forms: residence incorporates a complete series of villas, bungalows and high-rise buildings. In 2016, properties for sale have ushered in a number of new products such as enclosed courtyard, cascade compound and carefree yard, which are further addition and supplement to the original product types and welcomed by the market (see Fig. 4).

六、绿建及智能科技

2016年,万达商业规划牵头完成"慧云"(3.0版)

VI. GREEN BUILDING AND SMART TECHNOLOGY

In 2016, Wanda Commercial Planning took the lead in accomplishing R&D, testing, piloting, training and promotion of "Huiyun System V3.0", and initiated implementation of 181 Wanda Plaza "Huiyun System" annual new construction and renovation plans. In the

序号	时期	产品体系	代表项目	产品类别	品质分类	特点
1	2008年以前	明珠系列	长春明珠 大连明珠 南京明珠	别墅 洋房 小高 高层	中端	—
2	2008年至2015年	华府系列	大连华府 哈尔滨华府	洋房 小高 高层	中高端	配建酒店 万达广场
		公馆系列	武汉万达公馆 大连万达公馆 西安万达公馆	高层 超高层	高端	配建酒店 万达广场
		万达中心	武汉万达中心	高层 公寓	中高端	配建酒店
3	2016年	"院"系列	别院 版纳、成都、重庆、桂林 合院 成都、版纳、重庆、桂林 叠院 无锡、成都、重庆、桂林	别墅 合院 洋房	中高端	万达城

图4 销售类物业产品脉络

研发、测试、试点、培训及推广，牵头落实181个万达广场"慧云"年度新建及改造计划，将广场建设、运营与科技手段紧密结合，将万达广场由纯商业运营行为，提升到物联网、智能科技的高度，将现代科技与绿色建筑相结合。按照集团制度要求，所有万达广场都必须取得一星标准。截至2016年底，万达集团共计获得绿建认证488项，连续3年认证数量全国第一；其中2016年新增绿建认证129项，包括绿建设计标识89项，绿建运行标识25项，绿建饭店标识15项。2016年，万达集团还获得中国绿建委颁发的"全国绿色建筑先锋奖"；同时获得中国绿色建筑产业联盟颁发的"CIHAF2016年度绿色先锋企业"、"2016年度中国绿色地产运营10强"第一名等殊荣（如图5-7所示）。

七、万达BIM总发包管理模式研发

截至2016年底，万达BIM总发包管理模式研发已初见成效。此项研发涉及公司内部9大部门和外部23家研发单位，共计投入1315名研发人员，完成了全球首创的BIM"四大核心成果"。首先，基于万达多年商业开发经验，历经16版修订，建立了全球最领先的BIM标准模型，包括12大类专业信息、50万个标准构件和10亿条数据信息。目前已涵盖从方案到施工图全专业设计内容，覆盖全部建造标准，覆盖全国开发区域，奠定了产品标准化的基础。其次，开发出国际最先进的BIM功能插件，为开发项目管理提供了自动化、信息化辅助工具。第三、建立了全球首创跨企业BIM管理制度，构建了BIM多维度保障体系，实现了"四方"统一标准。第四，建立了全球首创的"四方"协同管理平台，统一平台实现"四方"协同工作，确保管理工作不受专业限制。同时，利用混合云构架实现了数据处理的高速安全。万达BIM总发包管理模式的研发，将BIM总发包管理模式提升到智能化管理高度，为成本降低和效率提升提供技术解决方案。2017年，BIM总发包管理模式将结合9个在施项目进行试运行，结合实际应用需求实现万达BIM总发包管理模式的不断迭代升级。

截至2016年底，万达新开业50个万达广场，2个万达茂，18个高端酒店（其中14家五星级酒店）。2016年，万达商业规划在人员编制缩减三分之一的情况下，共设计管控商业广场167个，销售物业133个，文旅项目14个，境外项目9个，酒店项目35个，每类项目数量均有20%~50%的增长。正是由于万达商业规划在标准版研发、业态创新、智能科技等创新方面不断取得进步，同时积极推进BIM总发包、设计总包等先进管控手段，才能如此大幅度降低成本、提升效率，实现上述不俗成绩。

图5 全国绿色建筑先锋奖奖牌

图6 中国年度绿色先锋企业奖牌

图7 中国绿色建筑TOP排行榜奖牌

process, plaza construction and operation are closely tied with technical means, Wanda Plaza is elevated to the level of Internet of Things and intelligent technology from a purely commercial business, and modern technology is integrated with green building. Wanda Plazas are required by the Group to obtain one-star Green Building label. At of the end of 2016, the Group had obtained a total of 488 "Green Building Certifications", ranking the first in terms of certification quantity for three consecutive years (of which, the Group obtained 129 "Green Building Certifications" in 2016, including 89 "Green Building Design Label" certifications, 25 "Green Building Operation Label" certifications and 15 "Green Hotel Label" certifications); In 2016, Wanda Group won the "National Green Building Leadership Award" issued by China Green Building Council, "CIHAF 2016 Green Leadership Enterprise" issued by China Green Building Association and ranked the first of the "Top 10 Chinese Green Real Estate Operators in 2016" (see Fig. 5-7).

VII. WANDA R&D ON BIM TURNKEY CONTRACT MANAGEMENT MODEL

As of the end of 2016, R&D of Wanda's Turnkey Contract Management Model had achieved initial success. A total of 1,315 staff from 9 internal departments and 23 external R & D units participated in the R&D work, attaining the world's first "four core achievements" in BIM. First, based on Wanda's years of business development experience and revision to 2016 version, the internationally-leading BIM standard model has been established, which contains 12 categories professional information, 500,000 standard components and 1 billion data information. At present, the BIM standard model has covered full-discipline design ranging from scheme to construction drawing, all the construction standards and the national development areas, laying foundation for product standardization. The most advanced functional plug-in in the world has been developed, which assists automation and information for development project management. Third, the world's first cross-enterprise BIM management system has been set up by means of constructing multi-dimensional BIM guarantee system and implementing "four-party" (i.e. Wanda Group, main design contractor, main project contractor and project supervisor) unified standard. Fourth, the world's first "four-party" collaborative management platform has been built. The platform enables collative work of four parties and ensures that the management work is not restricted by professionalism. At the same time, hybrid cloud architecture makes data processing high-speed and safe. Wanda R&D on BIM Turnkey Contract Management Model has upgraded the management model to an intelligent level, and offered technical solution for reduced cost and enhanced efficiency. In 2017, 9 projects under construction will try out the BIM Turnkey Contract Management Model, so that the management model could be continuously upgraded along with practical application requirements.

By the end of 2016, 50 Wanda Plazas, 2 Wanda Malls and 18 high-end hotels (14 of them are five-star hotels) had been newly opened. In 2016, with a one-third reduction of its staffs, Wanda Commercial Planning had designed totally 167 shopping plaza projects, 133 properties projects for sale, 14 cultural tourism projects, 9 overseas projects and 35 hotel projects, each category enjoying a 20% to 50% increase in total numbers. While actively promoting advanced management approaches (the BIM Turnkey Contract and General Design Contracting), Wanda Commercial Planning is unceasingly making progress in standardization R&D, business type innovation, intelligent technology and other innovations. Rightly with both efforts, we are able to significantly reduce costs, improve efficiency and attain the above achievements.

WANDA
COMMERC
PLANNI

万达城

WANDA CITY

WANDA CITY NANCHANG
南昌万达城

| 01

MASTER PLAN OF WANDA CITY NANCHANG
南昌万达城总体规划

PLANNING DURATION : MARCH, 2013 TO JUNE, 2016
LOCATION : NANCHANG, JIANGXI PROVINCE
PLANNED AREA : 290 HECTARES
FLOOR AREA : 4,960,000 SQUARE METERS

规划时间：	2013 / 03 – 2016 / 06
规划位置：	江西省 / 南昌市
规划面积：	290 公顷
建筑面积：	496 万平方米

一、缘起

"万达城"是万达集团凭借多年在商业、文化、旅游产业积累的丰富经验，整合全球资源，应用世界顶尖的管理经验和科技装备，结合独具特性的文化创意理念，倾力打造的"以游乐体验为主要特色"的特大型文化旅游商业综合项目。2013年至今，万达集团已在青岛、哈尔滨、南昌、合肥、无锡、广州、成都、桂林、重庆等城市开工建设十余个"万达城"。

二、南昌万达城选址

作为特大型文化旅游项目，选择合适的城市对"万达城"项目至关重要。选择南昌建设"万达城"，主要基于如下两点考虑。首先，"万达城"优先选择新一线城市及二线城市，南昌是江西省省会，属于二线城市，也是中部城市群中的佼佼者，经济处于高速发展的阶段，是"万达城"在中部区域布局的适宜之地。其次，"万达城"优先选择特色地域文化地区，南昌是国家"首批国家历史文化名城"，文化底蕴深厚，知名度高，这成为"万达城"落户的有力支撑。

南昌万达城于2016年5月18日正式开业，是万达集团在全国开业的第一个真正意义上的万达城，也标志着万达集团在文化旅游领域正式亮相。

1.用地条件

"万达城"的基地选址需要有两个必要的条件。首先，用地充足——南昌万达城的选址位于南昌红谷滩九龙湖新区，毗邻新的省级行政中心，可以满足万达城200~267公顷（3000~4000亩）净建设用地的要求（图1）。其次，区位优越，交通便利——九龙湖新区位于南昌城市发展方向上，近期主要的规划道路和市政设施建设能够通达；临南北向城市干道，又有规划地铁线到达；南昌万达城离昌北国际机场约40分钟车程，离南昌高铁西站约为20分钟车程，可有效承接内部和对外交通枢纽带来的大量客流。

2.旅游市场

"万达城"作为大型文化旅游项目，需要落户于旅游市场承载力强的区域。南昌作为中国优秀旅游城市，拥有世界文化遗产，国家级、省级及市级的重点文物保护单位50多处。每年南昌市以其独特的"红色、绿色，古色"景点吸引了一大批国内外游客前来观光旅游、投资兴业。南昌万达城文化复兴的项目定位对南昌整体旅游格局形成补充。通过"万达城"的建设，使南昌的文化旅游业更加繁荣。

I. ORIGIN

"Wanda City" is a super-large cultural, tourism and commercial mixed-use project brand presented by Wanda Group with its plentiful experience in the commercial, cultural and tourism industries. With "recreation experience as the main feature", Wanda City was constructed by integrating global resources, applying world leading management experience and sci-tech hardware in combination with unique cultural creative ideas. From 2013 to nowadays, Wanda Group has built or is building altogether more than ten "Wanda Cities" in different cities including Qingdao, Harbin, Nanchang, Hefei, Wuxi, Guangzhou, Chengdu, Guilin and Chongqing.

II. LOCATION OF WANDA CITY NANCHANG

As Wanda City is an extra-large cultural tourism project, its location becomes a crucial decision. Nanchang was selected mainly for the following two reasons. First, new first-tier cities and second-tier cities are the preferred places for constructing Wanda City. Nanchang, as the capital of Jiangxi Province and also an outstanding second-tier city in the city cluster of Central China, has a rapid-growing economy, making it an ideal home of

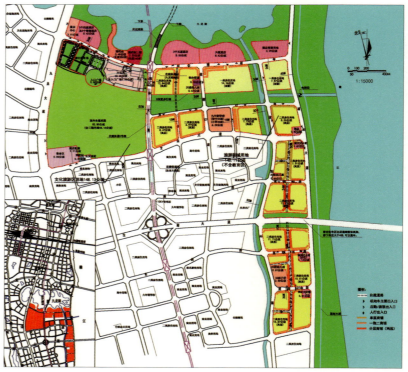

图1 南昌万达城规划图

图2 南昌万达城功能分区图

三、南昌万达城的规划策略

1.生态型规划布局

整体结构——规划结合周边的赣江以及九龙湖湿地公园进行生态化布局，形成"双轴引领，板块联动"的规划结构。"双轴"是指一条东西贯穿连接室外乐园以及赣江的绿化生态轴线；一条南北连通九龙湖湿地公园的蓝色生态轴线。规划以"双轴"为骨架，将核心区各功能地块进行连接，使整个区域形成一个有机的整体。

开敞空间——组团中布局开敞绿芯，由绿廊相互联通，结合"双轴"十字布局，使整个片区的绿化连成一体，营造出"点—线—面"空间序列。

景观体系——规划生态化的景观系统，并延续城市空间规划中"绿脉"的规划理念，保留"通湖达江"的绿色廊道；将室外主题公园融入城市中央景观带，把度假酒店纳入环九龙湖景观带，让区域的各个板块有机地融入周边的生态环境中。

2.复合型的功能构成

聚集性——南昌万达城占地面积290公顷，总建筑面积496万平方米，包含万达茂、室外主题乐园、度假酒店区和生态新城"四大功能"板块。其中万达茂占地面积20公顷，度假酒店区占地35公顷，室外主题乐园占地面积80公顷。"四大板块"聚集在万达城内，形成一个巨大的文化旅游度假区（图2）。

复合性——在南昌万达城的建设中，整合了室外主题娱乐、室内主题娱乐、酒店度假、商业商务、生态居住等多功能的复合业态，让南昌万达城成为一个内涵丰富、多元化构成的旅游度假区，与万达集团综合发展的企业理念相契合。

融合性——根据不同业态的使用和运营要求，进行合理布局；通过规划游线系统，将各个功能板块进行串联，并相互带动，让"万达城"成为一个相互融合的有机体。

3.和谐的空间秩序

总体布局——南昌万达城的总体布局采用尊重自然环境的和谐布局策略，结合赣江和九龙湖生态湿地，建筑布局与生态界面"共生共存"——靠近赣江和九龙湖一侧的布局低层及多层建筑；远离赣江和九龙湖湿地的区域，建筑逐渐升高，呈阶梯状递进式布局。

天际线规划——南昌万达城在空间设计上充分考虑到天际线的变化，规划建筑高

Wanda City in Central China. Secondly, distinctive geography and culture featured regions are the preference of Wanda City. Nanchang, as one of the "First National Historical and Cultural Cities", enjoys a profound culture and wide popularity, making it a perfect place for settling Wanda City.

Wanda City Nanchang was launched officially on May 18, 2016. It is the first Wanda City formally opened in China, marking the debut of Wanda Group in the cultural and tourism field.

1. LAND USE CONDITIONS

The location of "Wanda City" was selected with two basic conditions. First of all, the land shall be sufficient - Wanda City Nanchang is located in the Jiulong Lake New Area, Honggutan in Nanchang, adjacent to the new provincial administrative center, with a net developable land up to the Wanda City's requirement, that is 200 to 267 hectares. (3,000 to 4,000 acres) (Fig.1). Secondly, the location shall be advantageous with convenient transportation - Jiulong Lake New Area is situated at the urban development direction of Nanchang, and could be accessed by the major roads and municipal facility construction planned recently; it nears the north-south city trunk roads and is connected by planned subway lines; it takes about 40 minutes from Nanchang Wanda City to Changbei International Airport and about 20 minutes to the West Nanchang High-speed Railway Station. These enable it to efficiently afford the huge passenger flows coming from internal and external transport hubs.

2. TOURISM MARKET

As a large-sized cultural tourism project, Wanda City shall be located in an area with strong tourism market bearing capacity. Nanchang, as one of the excellent tourism cities in China, has more than 50 key cultural relics under world-class, national, provincial and municipal protection. Every year, a large number of domestic and foreign tourists, attracted by Nanchan's unique "red, green and ancient" scenic spots, come here for sightseeing and investment. The orientation of Wanda City Nanchang as a cultural revitalizing project will reinforce the overall tourism development of Nanchang. The construction of Wanda City will promote the cultural tourism of Nanchang.

III. PLANNING STRATEGIES FOR WANDA CITY NANCHANG

1. ECO-PLANNING LAYOUT

Overall structure - the planning aims to form an ecological layout in combination with the Ganjiang River and Jiulong Lake Wetland Park in the surrounding environment, so as to build a planning structure "led by dual axes, with districts linked with each other". "Dual axes" refers to the east-west running-through green eco-axis that connects the outdoor park with the Ganjiang River, and the north-south blue eco-axis that connects to the Jiulong Lake Wetland Park. In the planning, the "dual axes" act as the framework linking different functional plots of the core area, so that the entire area becomes an organic wholeness.

Open space - in the cluster's layout "green cores" are opened and interconnected via the "green corridor". In conjunction with the "dual axes" cross layout, afforestation throughout the zone is joined into an integral whole, creating a "point-line-surface" spatial sequence.

Landscape system - an ecological landscape system is planned by carrying on the "green

图3 南昌万达城鸟瞰图

低有致、前后错落；在整体上形成优雅动人、变化丰富的天际轮廓线，如图3所示。

4.一体化的交通体系
南昌万达城规划整合周边交通资源，深化内部交通系统，使万达城形成内外一体化的交通格局。

道路系统规划——依托现有的九龙大道、生米大道、城运大道、国体大道和生米南大道，形成万达城内的主要道路交通骨架。在干路系统基础上，布置支路网系统，形成主次支三级路网等级。按照各道路的"交通型"与"生活型"功能需求划分断面形式。做到："交通型"道路断面——更加强调货运道路的通过性，双向车道不干扰；"生活型"道路断面——更加强调慢行交通和行人的舒适性和安全性。

公交系统规划——结合南昌地铁2号线设置两处站点，分别为"九龙湖南站"和"生米大道站"。其中"九龙湖南站"位于万达茂、酒店、酒吧街和生态居住区的功能板块中间，能更大范围地为万达城提供服务；结合九龙大道、城运大道、环湖路和生米南大道设置公交线路及站点，站点间距控制在500~800米范围内；并结合万达茂及室外主题乐园出入口设置出租车停靠站，作为公交系统的有益补充。

慢行系统规划——"万达城"内部规划了多样化的慢行系统。以万达茂为核心，形成"体验型"的慢行系统。在这种"一体化"的空间体系中，步行系统串联办公、购物、娱乐、餐饮、聚会等人群活动形成互相支持的、便捷、舒适、流畅的交通系统；以"双轴"为核心，形成"生态型"的慢行系统。在这种"生态型"的慢行空间中，赣江、九龙湖生态湿地公园等"生态化斑块"与"万达城"融为一体，为休闲"慢生活"提供载体。

network" concept in urban space planning and preserving the green corridor that "accesses the lake and connects the river"; the outdoor theme Park is integrated into the city's central landscape belt, and the resort hotel into the landscape belt around Jiulong Lake, so that all blocks of the area becomes organic parts of the surrounding ecological environment (Fig.3).

2. COMPOSITE FUNCTIONAL CONSTITUTION
Aggregation - Wanda City Nanchang covers an area of 290 hectares and a gross floor area of 4,960,000 square meters. It consists of four major functional zones / districts, namely Wanda Mall, outdoor theme park, resort hotel zone and ecological new city. Among them, Wanda Mall occupies 20 hectares, the resort hotel zone 35 hectares and the outdoor theme park 80 hectares. Gathering in Wanda City, the "four major parts" form a huge cultural tourism resort (Fig.2).

Complexity - in the construction of Wanda City Nanchang, complex formats are integrated to provide multiple functions including outdoor theme entertainment, indoor theme entertainment, hotel vacationing, commercial and business, ecological residence and so on. These facilities make Nanchang Wanda City a tourist resort with profound connotations and diversified constitutive elements, which is consistent with Wanda Group's development philosophy.

Integration - a reasonable layout is made according to the different using and operational requirements of different business types; by planning the tourist route system, various functional parts are connected in series so that they could drive each other, building Wanda City into an organic body with mutual integration.

3. HARMONIOUS SPATIAL ORDER
Overall layout - the overall layout of Wanda City Nanchang is made following the principle "respecting the natural environment for harmony". Based on the surrounding conditions of the Ganjiang River and Jiulong Lake ecological wetland, an "ecological coexistence" construction layout is arranged. To be specific, at the side close to the Ganjiang River and Jiulong Lake, low-rise and multi-storey buildings are mainly designed; in the area far away from the Ganjiang River and Jiulong Lake wetland, higher buildings are arranged in a "the farer, the higher" layout.

Skyline planning - the changes of skyline is considered as an important spatial design element for Wanda City Nanchang. Buildings are of different heights and distributed properly, so as to present an irregular, undulating, variable while well-portioned skyline effect.

4. INTEGRATED TRANSPORTATION SYSTEM

The planning of Wanda City Nanchang, by integrating transportation resources of the surrounding area, making in-depth improvement of the internal transport system, provides an integrated traffic framework incorporating both inside and outside the Wanda City.

Road Network - the main road traffic framework within the Wanda City is shaped with the support of existing Jiulong Avenue, Shengmi Avenue, Chengyun Avenue, Guoti Avenue and South Shengmi Avenue. Based on the trunk road system, branch network systems is arranged to form a "main-secondary-branch" three-level road network. Cross-sectional forms are selected according to the functional needs of different roads, which could be "traffic use", "life use" and so on. The cross-section of "traffic use" roads shall guarantee going-through for freight purposes, with both driveways not interfering each other; for "life use" roads, more emphasis shall be laid on slow traffic and pedestrian comfort and safety.

Public Transportation Network - two stations are planned considering Line 2 of Nanchang Subway, namely "the South Jiulong Lake Station" and "the Shengmi Avenue Station". The South Jiulong Lake Station sits in the middle of the functional blocks of Wanda Mall, hotels, bar street and ecological residential area, and therefore could serve a large range in Wanda City; bus lines and stations are arranged considering the situations of the Jiulong Avenue, Shengmi Avenue, Chengyun Avenue, Guoti Avenue and South Shengmi Avenue; inter-station distances are controlled at 500m to 800m; at entrance/exit of Wanda Mall and the theme park taxi stops are designed as supplement to the bus system.

Slow Mode - a diversified slow traffic system is designed within Wanda City. This will be a slow traffic system of "experience" with Wanda Mall as the core area. In such an "integrated" spatial system, the pedestrian network connects office, shopping, entertainment, dining, gathering and other people activities, forming a mutually-supportive, convenient, comfortable and smooth transportation system; with "dual-axes" as the center, a slow traffic system come into being. In this "ecological type" slow traffic space, the Ganjiang River, Jiulong Lake Ecological Wetland Park as well as other "ecological spots" fuse into Wanda City to provide a casual "slow life" together.

5.CONTINUAL CULTURAL HERITAGE

Culture carries the spirits of a region. Wanda City does a lot to excavate regional cultures. Nanchang, being a time-honored city with a history of thousands of years, has its unique customs and distinctive cultures. In the ancient times, Nanchang enjoyed a very high reputation in making celadon, producing lacquer and textiles, bronze ware manufacture and production process of gold, silver and jewelries. It had been metallurgy, textile, shipbuilding center and commercial city in the river south. In modern Chinese history, Nanchang is also a city of heroes with glorious revolutionary traditions. "Red Culture" has also become one of the features of Nanchang.

The planning of Wanda City digs deep into unique local history and folk cultures, makes refinements and innovations and applies them in the architectural forms, play ideas, theme packaging and artistic performances of Wanda City, so that it becomes a culture carrier that is highly "participatory and interactive". The landscape design of Wanda City adopts the cultural essence of Chinese River South style gardens, with the artistic conception of "small bridge over flowing stream" and "winding path leading a seduced spot".

The "intangible cultural heritages" of Jiangxi Province are also applied in the planning of Wanda City Nanchang, including major landscape features, traditional performing arts, landmark buildings, famous historical figures, myths and legends, etc. Modern technologies, packaging forms and expression techniques are also employed to present the history and promote local cultures through "theme creation".

IV. CONCLUSION

The planning of Wanda City Nanchang features "being consistent with the existing development orientation of Nanchang city, fully respect the geographical features, perfectly integrate with the general urban landscape framework program", and will accomplish "win-win" via mutual growth of the Wanda City and Nanchang city. The grand opening of Wanda City Nanchang further enhances the brand effect of "Wanda Cultural Tourism" and attracts the attention of the whole world!

5.延续性的文化传承

文化是一个地区的精神载体，因此"万达城"极其注重地域文化的挖掘。南昌是一个历史悠久的城市，有几千年历史，有着独特鲜明的风俗习惯和特色文化。古代南昌在青瓷器的烧造，漆器、纺织品的生产，铜器制造以及金、银首饰品的生产工艺方面都有很高的声誉，这里一度是江南的冶炼、纺织、造船的中心和商业都市。在近代历史中，南昌又是一座富有光荣革命传统的英雄城市，"红色文化"也成为南昌的特色文化。

在"万达城"的规划中，深入挖掘地方特有的历史及民俗文化，并对其进行提炼和创新，运用到"万达城"的建筑造型、节目创意、主题包装、文艺表演之中，使"万达城"成为一个"参与互动性"极强的文化载体。在景观设计上，突出中国江南园林的精髓，以"小桥流水"、"曲径通幽"的意境，突出中国园林文化。

南昌万达城也将江西的"非物质文化遗产"资源运用到规划中，包括重要地物地貌、传统表演艺术、地标建筑、著名历史人物、神话传说等，利用现代技术、包装形式和表现手法，在"主题营造"中传承历史，弘扬当地文化。

四、结语

南昌万达城规划具有"与现有南昌城市发展方向相契合、充分地尊重地域特征、完美地与城市景观大格局项目融合"的规划特点，从而实现"万达城"与城市协同发展的"双赢"局面。南昌万达城的全面开业盛况，使得"万达文化旅游"的品牌效应更加凸显，让世界瞩目！

02
PUBLIC LANDSCAPE PLANNING OF WANDA CITY NANCHANG
南昌万达城公共景观规划

OPENED ON: 300,000 HECTARES
PLANNING PHILOSOPHY: LOW-CARBON, ENVIRONMENTALLY FRIENDLY, AND HEALTHY LIVING
DESIGN TIME: 2012 TO 2014
LOCATION: NANCHANG, JIANGXI PROVINCE

项目规模： 30万公顷
规划理念： 低碳环保，健康生活
设计时间： 2012 – 2014
项目地点： 江西省南昌市

一、项目概况

南昌万达城公共景观代表性的规划项目是南昌万达城中央公园。中央公园坐落于红谷滩新区九龙湖片区，占地总面积30公顷，是南昌万达城园林绿地系统中重要的组成部分之一，也是万达城住宅区的核心部分，为周边市民提供阳光舒适、休闲放松的生态健康休闲场地。同时串联九龙湖上下游，完善环九龙湖区域的绿化体系，形成生态低耗、功能完善的绿色廊道，满足各种类型的使用功能需求，塑造各种不同私密程度的开放空间，带领万达城居民享受"慢生活"和"高品质"。结合南昌当地鲜明的传统青花瓷文化，打造"沐润苍翠"、"九曲花溪"、"烂漫花海"、"刺绣迷宫"、"碧水红枫"、"纸鸢青草"、"青瓷雅苑"、"青瓷童趣"、"水涧花溪"和"赏水栈台"等十大景观节点。

二、规划理念

南昌万达城中央公园本着"低碳环保、健康生活"的设计理念，积极保护和改善城市生态环境，大力推动南昌市生态文明建设，打造南昌万达城"海绵城市"。同时，以人为本，服务于健康生活，为市民提供丰富多样的室外运动场地，并对地域文化元素的进行提炼，运用了大量当地材料如青花瓷、青砖、青瓦等，以精湛工艺的继承，在城市中央打造独具南昌特色的兼顾人文性、闲适性、自然性城市绿肺（图1）。

三、规划手法

1."低碳环保"——生态文明的海绵城市

南昌万达城中央公园采用了浅层调蓄、低影响开发的模式，其旨在尽可能从源头上处理好水资源，减少对原有水文的破坏，也即从水的角度来指导城市的设计和开发。并通过自然存积、自然渗透、自然净化等功能，打造在适应环境变化和应对自然灾害等方面具有良好弹性的"海绵城市"（图2）。

中央公园东西两地块的大草坪作为生态滞留区形成浅水洼地及景观区，是利用工程土壤和植被来存储和治理径流的一种形式，治理区域包括草地过滤、砂层和水洼面积、有机层或覆盖层、种植土壤和植被。通过将雨水滞留下渗来补充地下水并降低暴雨地表径流的洪峰，其中浅坑部分能够蓄积一定的雨水，延缓雨水汇集的时间，土壤能够增加雨水下渗，缓解地表积水现象。蓄积的雨水能够供给植物利用，减少绿地的灌溉水量。

图1 中央公园鸟瞰图

I. PROJECT OVERVIEW

The Central Park of Wanda City Nanchang is a representative public landscape planning project of Wanda City Nanchang. Located in the Jiulong Lake area, Honggutan New District, the Central Park covers a gross land area of 30ha. It is a significant component of the green space system of Wanda City Nanchang, as well as a core part of Wanda City residential area, which provides the citizens with an ecological venue for health preservation, relaxation and entertainment. The Park connects to both the upstream and downstream of Jiulong Lake, forming a perfect green system for the area surrounding the Lake with an ecological, low-consumption green corridor with complete functions. It can meet various types of requirements on usage functions, create open spaces of different privacy degrees, and offer people in Wanda City with "slow life" and "high quality". In combination with the distinctive traditional local "blue and white porcelain" culture, ten major landscape nodes are constructed, namely the "Verdant Gully", "Huaxi Brook with Nine Bends", "Romantic Sea of Flowers", "Embroidered Labyrinth", "Green Water with Red Maple", "Kite with Green Grass", "Celadon Jade Palace", "Celadon Childhood", "Water Gully and Huaxi Brook" as well as "Water-viewing Stack Table".

II. PLANNING PHILOSOPHY

Adopting the design philosophy "low-carbon and environmentally friendly, and healthy living life", the Central Park of Wanda City Nanchang aims to actively protect and improve the urban ecological environment, vigorously promote the ecological civilization construction of Nanchang, and build a "Sponge City" of Wanda City Nanchang. With the principle "people-oriented, and service healthy living", it offers the public with a variety of outdoor sports spaces, where a large number of local materials such as blue and white porcelain,

图2 入口广场

图3 生态透水铺装

图4 活动广场

图5 儿童活动场地

园区内大量运用透水铺装，加大雨水渗透量，减少地表径流，渗透的雨水储蓄在地下储蓄池内经净化排入河道或者补给地下水，减少了直接性雨水对路面冲刷然后快速径流排水对于水源的污染。

项目中还设计了多处雨水塘，周边种植耐湿植物，形成低矮绿篱，消除安全隐患。同时整个雨水塘系统形成微循环，补给地下水的同时有效地削减径流峰值。

植栽方面：以自然速生的当地植栽为主，可快速搭建起整个园区的绿量。同时近人尺度多采用开花乔灌木，表现四时有绿四季有花的植栽意象，对该地区的生态恢复及植物多样性展示具有重要意义。

2."健康生活"——丰富多样的运动空间

南昌万达城中央公园设计有两条生态运动带：一条是5000米的环形自行车道，一条为近10000米的健康运动慢跑道。跑道上分别设计了长度提示，起点终点提示及运动类别提示，方便市民选择及引导。

为满足周边市民不同的需要，中央公园提供充足的、多样的运动场地，包括篮球场、羽毛球场、乒乓球场等。同时，也满足了不同年龄段人们的需求，分别设有老年活动广场、儿童活动广场，其中老年活动广场除了满足广场舞的场地需求（图4），也有丰富的老年健身器械；儿童活动广场，分为3~5岁、5~7岁及8~12岁三个年龄段，全塑胶场地铺装保证了儿童活动安全需求，给儿童多彩且文化特色浓郁的活动空间，以及丰富的儿童游戏装置（图5）。

grey bricks and gry tiles are used carrying regional cultural elements. By inheriting the superb, exquisite crafts and techniques, it is built into the green lung at city center that presents both local characteristics and such general features as humanity, relaxation, and naturalness (Fig.1).

III.PLANNING APPROACH

1. LOW-CARBON AND ENVIRONMENTALLY FRIENDLY - "SPONGE CITY" OF ECOLOGICAL CIVILIZATION

The Central Park of Wanda City Nanchang adopts shallow regulation and storage as well as low-impact development, so as to process water resources properly from the sources as far as possible, and to cause as less as possible damages to the original hydrology. In other words, urban design and development is made from the perspective of water conservation. In addition, natural accumulation, natural infiltration, natural purification and other functions are utilized to build a "Sponge City" that is flexible in adapting to environmental changes and responding to natural disasters (Fig.2, Fig.3).

The large lawns at the east and west plots in the Central Park, as ecological retention areas, form shallow lowlands and landscape zones, aim to store and govern rainwater runoffs by use of engineering soil and vegetation. The governed coverage are grass filtering, sand stratum and water puddle area, organic layer or overburden layer, cultivated soil and vegetation. It retains stormwater so that it infiltrates into the ground, making up groundwater and reducing the flood peak of storm runoffs; the shallow pits could store certain amount of accumulated rainwater, and thus delaying rainwater gathering time; the soil can increase rain water infiltration and therefore mitigates surface water catchment The accumulated rainwater can be supplied to plants, reducing the water needed for greenbelt irrigation.

Permeable pavement is largely applied in the Park, so as to increase rainwater infiltration and reduce surface runoffs. Infiltrated rainwater is stored in the underground storage pool, and discharged into river courses or replenishes groundwater after being purified, so as to reduce the pollution caused by rapid runoff drainage from direct rainwater wash-out on road surface.

A number of stormwater ponds are designed for the Project, with damp tolerant plants growing in the surrounding areas, forming low hedgerows that could eliminate safety hazards. Moreover, micro-circulation of the entire stormwater pond system could not only make up groundwater, but also effectively put down the runoff peak value.

Planting: mainly with naturally fast-growing local plants, so that vegetation quantity of the entire Park could be achieved in a short time. At the same time, flowering trees and shrubs are mainly planted in areas close to persons to present the ever-green and ever-flowering imagery. This is significant for the ecological restoration and plant diversity presentation of the area.

2.HEALTHY LIVING - RICH AND COLORFUL SPORTS SPACES

Two ecological sports belts are designed for the Central Park of Wanda City Nanchang. One is the 5000m annular bikeway, and the other is a health sports jogging track nearly 10,000m. On the jogging track there are length signs, starting/terminal point signs and sports type signs for convenience.

To meet different needs of the public, adequate, different sports venues are provided in the Central Park, including basketball courts, badminton courts, table tennis courts, etc. Demands of different ages are also taken into consideration. There are activity fields for senior citizens, children's activities field, etc. The fields for the aged can be used for square dancing, with a number of fitness facilities for the elderly also installed (Fig.4). Children's playgrounds are divided into three types, respectively for age groups from 3-5, 5-7 and 8-12; all-plastic ground guarantees safety, and the spaces are arranged with rich colors and cultural characteristics, as well as a wealth of children game devices (Fig.5).

03

WANDA MALL NANCHANG
南昌万达茂

OPENED ON: 28th MAY, 2016
LOCATION: JIULONG LAKE NEW AREA, NANCHANG
LAND AREA: 22.05 HECTARES
FLOOR AREA: 192,800 SQUARE METERS

开业时间：2016 / 05 / 28
开业地点：南昌市 / 九龙湖新区
占地面积：22.05 公顷
建筑面积：19.28 万平方米

万达茂概述

南昌万达茂位于南昌九龙湖新区万达文化旅游城地块，毗邻九龙湖及赣江，区域优势明显，包含万达广场、海洋乐园、电影乐园"三大业态"，给人们提供多方位的休闲娱乐体验。南昌万达茂西靠南龙蟠路，东临三清山大道，南侧是万达主题乐园，西侧为大型地面停车场及入口公用广场；地铁二号线经停万达茂2号门外广场，交通便利。万达茂商业建筑由购物中心、室外步行街及甲级写字楼组成。其中大商业建筑面积13.98万平方米，海洋乐园建筑面积3.68万平方米，电影乐园建筑面积1.57万平方米，VIP用房500平方米。

PROJECT OVERVIEW

Wanda Mall Nanchang has a prestigious location in the Wanda cultural tourism plot in Jiulong Lake New Area, Nanchang city, adjacent to the Jiulong Lake and Ganjiang River. It comprises of "three major programmes" respectively being Wanda Plaza, Ocean Park and Movie Park to provide people with all-round leisure and entertainment experience. Wanda Mall Nanchang borders the Longpan Road in the west and Sanqingshan Avenue in the east; in its south part is Wanda Theme Park, and in the west side are the surface car park lot and entrance to the Public Square; it has convenient transportation with Metro Line 2 station on the square outside No. 2 gate of Wanda Mall. Wanda Plaza consists of the shopping mall, outdoor pedestrian street and class A office buildings. Among them the floor area for commercial function is 139,800 square meters, the Ocean Park's floor area is 36,800 square meters, and floor area of Movie Park and the VIP rooms are 15,700 square meters and 500 square meters respectively.

1 南昌万达茂总平面图
2 南昌万达茂全景图
3 青花瓷纹样手绘图
4 南昌万达茂BIM板材优化归类1
5 南昌万达茂BIM板材优化归类2

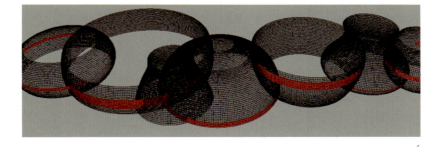

4

5

设计管控

南昌万达茂位于被誉为"瓷都"的江西。规划院团队以"青花瓷"作为主要元素，深入挖掘当地文化内涵，定标"蕴育青花"、"吉祥如意"的主题概念，体现江西"千年瓷都"的风貌。南昌万达茂立面设计取意"青花瓷"，最初设想立面材料均为瓷板，但在方案"深化阶段"和"扩初阶段"，通过多轮现场材料样板效果研究和动态成本测算，最终确立了彩色打印瓷板、双色喷漆铝板及涂料的组合方案，在确保主立面效果的同时有效控制了成本。南昌万达茂的双曲面幕墙由4.5万片瓷板组成，其规格超过一千种，每片瓷板的尺寸、形状和花纹都是不同的。通过BIM技术的运用，有效减少了面板规格，降低了生产成本，提高了实施效率；同时采用多色阶数字化无损处理技术，完美呈现青花瓷的韵味。伴随南昌万达茂的盛大开业，其曼妙的曲线、精彩的意象，宛若一件巨大的雕塑艺术作品，令全世界的目光汇聚于此，感受"青花瓷"独特的艺术光芒。

DESIGN CONTROL

Wanda Mall Nanchang is located in Jiangxi Province which is known as the "Porcelain Capital". Therefore, the design institute adopts "Blue and White Porcelain" as the main concept in its design and deeply explored the local cultural context. Then "Cultivation of Blue and White Porcelain" and "Good Fortune and Luck" are defined as the theme concept to reflect the features of Jiangxi as "Thousand-year Porcelain Capital". In accordance with the elevation design theme "Blue and White Porcelain", the preliminary attempt was to use porcelain plate as the elevation material. However through several rounds of plot material model effect studies and dynamic cost calculations during the "design development" and "expanded preliminary design" phase, the final scheme was made up to adopt a combination of color print porcelain plates, double-color painted aluminum plates plus coating, which could ensure the effect of front elevation while effectively control the costs at the same time. The double-curved surface curtain wall of Wanda Mall Nanchang is made up of 45,000 pieces of porcelain plates in over one thousand categories, each piece being of different sizes, shapes and patterns on them. The use of BIM technology effectively reduced the panel specifications, reduced production costs and improved implementation efficiency; the multi-gradation digital non-destructive processing techniques perfectly presented the charms of blue and white porcelain. As Wanda Mall Nanchang made its grand appearance, it attracts the eyesight of the whole world like a giant sculpture with its graceful outlines and wonderful imagery, radiating the unique artistic glows of "Blue and White Porcelain".

6 南昌万达茂鸟瞰图

7

万达茂外装

万达茂自筹建之初，便秉承将其打造为南昌乃至江西的"文化名片"、"标志性地域文化建筑"的理念进行构思设计。外观由英国STUFISH公司完成，选择了代表"江西文化特征"的瓷器作为外观设计元素。鸟瞰万达茂是26个形态各异的罐子紧密地排列在一起；从地面不同角度观赏，则可以看到优美曲线的各种罐体和丰富的青花彩绘图案。南昌万达茂不仅仅是一个融合商业、文化、娱乐属性的综合体建筑，设计师更希望将其打造成一座"建筑雕塑"。

FACADE DESIGN

Since its preparation, Wanda Mall undertook the mission to become the "Cultural Business Card" and "Iconic Regional Cultural Architecture" of Nanchang and even Jiangxi Province; this has led its concept design. The appearance was designed by world famous British company STUFISH, with porcelain which represents the cultural characteristics of Jiangxi as the main design element. From the bird's-eye view Wanda Mall is 26 different jars of different shapes tightly arranged together; from different angles on the ground, you can see the beautiful contours of the jars and rich blue and white painted patterns on them. Not merely a complex architecture integrates commerce, culture and entertainment, the designer expected more to build Wanda Mall Nanchang into a an "architectural sculpture".

8

PART B WANDA CITY
万达城

031

9

7 南昌万达茂2号入口
8 南昌万达茂立面图
9 南昌万达茂1号入口图
10 南昌万达茂外立面

10

PART B WANDA CITY 万达城

万达茂内装

内装设计融入对南昌地域文化的思考，让江西千年"瓷文化"再次绽放异彩；同时将建筑外装的语言引进室内，做到室内外相统一。青花瓷，白如玉、薄如纸、明如镜、声如磬。将青花瓷的"形"、"色"、"品"、"性"，融入建筑室内空间当中，做到传统和现代的统一。

INTERIOR DESIGN

The interior design incorporates thinking about the geographical culture of Nanchang, reviving the thousand-year "Porcelain Culture" of Jiangxi; at the same time, it brings the concepts of exterior decoration indoors and thus achieves inside-out unity. The blue and white porcelain features being white as jade, thin as paper, bright as mirror and sound as chime. Such characteristics in in "shape", "color", "quality" and "nature" are fused into the architecture, accomplishing the unification of tradition and modernity.

11 南昌万达茂中庭
12 南昌万达茂室内步行街
13 南昌万达茂室内步行街侧裙
14 南昌万达茂室内步行街连桥
15 南昌万达茂商业落位图
16 南昌万达茂室内步行街

14

15

16

17

18

万达茂景观

南昌万达茂位于南昌万达城主轴线上，景观面积为5万平方米，分为主次入口广场、带型广场及后勤区。按景观节点序列，以具象或抽象的手法展示传统的制瓷工艺，通过"拉胚"、"旋胚"、"画胚"、"施釉"和"烧窑"这"五道主要工序"形成"五大景观主题节点"，与青花瓷建筑群互相呼应，传达"青花瓷文化"。"圆形"被确立为设计的"平面构图母体"，其源于建筑外立面为瓷罐之形，是对圆弧形建筑轮廓的延续和深化。在此框架下，引入弧形铺装、弧形带状绿化、弧形及圆形水景池等景观元素。

LANDSCAPE DESIGN

Located on the main axis of Wanda City Nanchang, Wanda Mall Nanchang covers a landscape area of 50,000 square meters, divided into the primary and secondary entrance squares, linear plaza and service area. In a sequence of landscape nodes, the traditional porcelain producing technologies are presented in a figurative or abstract manner. "Five major landscape theme nodes" correspond to "five main manufacturing procedures" including "pulling rough", "rotating rough", "drawing on rough", "glazing" and "kilning", echoing the blue and white porcelain building cluster and conveying the "Blue and White Porcelain". "Circular" was basic of the "plane composition" because the buildings' facades are in the shape of porcelain jars. It continues and deepens the circular architecture contours. In such a framework, landscape elements like curved pavement, curved strip-shaped planting, curved and circular waterscape pools and so on are employed.

17 南昌万达茂广场喷泉
18 南昌万达茂景观小品
19 南昌万达茂水景
20 南昌万达茂景观
21 南昌万达茂景观小品
22 南昌万达茂夜景

万达茂夜景

夜景照明也要再现"青花瓷"这种典型的中国艺术形式。因此，将"晕染之光，玲珑之光"作为设计的主题思想——通过光线的变化和转移，在体现建筑之美的同时，赋予万达茂丰富的夜间表情和新的生命力。通过灯具变焦实现灯光晕染流动的效果，将光的柔美与建筑主体相结合，使得万达茂如同一组巨大的蓝色夜光瓷瓶在瓷都的夜幕下交叠辉映，美不胜收！

NIGHTSCAPE DESIGN

The nightscape lighting shall also reproduce the classical "Blue and White Porcelain" Chinese art form. Therefore, the main design idea "blending light and delicate light", via the changes and transfer of lights, not only reflects the beauty of architecture but also empowers Wanda Mall with diversified night faces and new vitality. Lighting zooming is utilized for light fading and flowing, combing the grace of lights with the main architecture structures, so that Wanda Mall, like a series of giant blue luminous porcelain vase, overlapping each other and shining in the night, creating an incredibly beautiful scene.

04

HOTEL COMPLEX OF WANDA CITY NANCHANG
南昌万达城酒店群

OPENED ON : 28th MAY, 2016
LOCATION : WANDA CITY NANCHANG, NANCHANG
LAND AREA : 25.88 HECTARES
FLOOR AREA : 206,700 SQUARE METERS
LANDSCAPE AREA : 205,200 SQUARE METERS

开业时间：2016 / 05 / 28
开业地点：南昌市 / 南昌万达城
占地面积：25.88 公顷
建筑面积：20.67 万平方米
景观面积：20.52 万平方米

1

酒店群概况

2016年5月28日，南昌万达城酒店群盛大开业。酒店群项目位于南昌万达文化旅游城内，东临赣江，南靠南龙蟠街，西接九龙大道，北依九龙湖。酒店群规划有8座度假酒店及酒吧街。结合紧邻九龙湖湖岸的优越的自然条件，在概念设计中酒店群的整体风格考虑为度假休闲型酒店。沿九龙湖湖岸从东往西依次展开南昌万达文华度假酒店（六星）、南昌万达嘉华度假酒店（五星）、南昌万达铂尔曼酒店（五星）、酒吧街、南昌万达诺富特酒店（四星）、南昌万达美居酒店（四星）及三座三星级主题销售酒店。

南昌万达城酒店群环九龙湖而建，依地形居高临水，远眺湖面，"浩浩荡荡，横无际涯，上下天光，一碧万顷"，可谓蔚然大观。在规划格局上以规模效应和互补完善的功能形成山水之外的"大势"。 各个酒店围绕九龙湖特点展开，有"八大主题景观"，突出"九彩湖畔，异域风情"的整体景观理念。酒店群规划"一"字排开，独享3千米长超大观湖空间。酒店客房内可远观壮丽湖景，也可近距离体验滨湖设置的主题景观。赏景同时，提供环整个九龙湖的10千米慢跑道，赏景与健身融为一体。

PROJECT OVERVIEW

On May 28th, 2016, Wanda City Nanchang Hotel Complex opened to the public. Situated in Nanchang Wanda Cultural Tourism City, the Hotel Complex adjoins the Ganjiang River in the east, approaches the South Longpan Street in the south, connects to the Jiulong Avenue to the west, and borders on the Jiulong Lake to the north. Eight resort hotels and a bar street are planned for the Hotel Complex. Based on the superior natural conditions close to the bank of Jiulong Lake, the conceptual design defines Hotel Complex as a place for resort and vacationing. From the east to the west along the lakeside of Jiulong Lake, are Wanda Vista Resort Nanchang (6-star), Wanda Realm Resort Nanchang (5-star), Pullman Nanchang Wanda (5-star), Bar Street, Novotel Nanchang Wanda (4-star), Mercure Nanchang Wanda (4-star) and three 3-star hotels for sale.

The Hotel Complex, built around the Jiulong Lake, overlooks the lake from above, "going forward with great strength and vigour in magnificence and interminability, the water and sky fusing into oneness, watery blue reaching far beyond the horizon", presenting an eyesight-attracting, grand sight. In the planned layout, scale effects and mutually complementary functions are utilized to form the "grand gesture" on the basis of landscapes. The hotels are designed centering on the features of the lake. With specialtites like "the eight major theme landscapes", the overall landscaping idea of "multi-colored lakeside with exotic feelings" is highlighted. The Hotels planned in a single-line formation, with an exclusive 3km long ultra-large space, view the lake scenes. Guests can overlook the magnificent lake views in the hotel rooms from afar, or get close to experience the theme landscapes at lakeside. While apprieciating the beautiful scenes, they can do physical excercises on the 10km Jogging trail around the whole lake.

PART B WANDA CITY
万达城

037

1 南昌万达城酒店群鸟瞰图
2 南昌万达城酒店群总平面图

WANDA VISTA RESORT NANCHANG
南昌万达文华度假酒店

OPENED ON: 28th MAY, 2016
LOCATION: WANDA CITY NANCHANG, NANCHANG
LAND AREA: 8.18 HECTARES
FLOOR AREA: 45,700 SQUARE METERS
LANDSCAPE AREA: 70,400 SQUARE METERS

开业时间：2016 / 05 / 28
开业地点：南昌市 / 南昌万达城
占地面积：8.18 公顷
建筑面积：4.57 万平方米
景观面积：7.04 万平方米

酒店概况

南昌万达文华度假酒店为六星级豪华酒店，占地8.18公顷，总建筑面积4.57万平方米，拥有客房203间。

PROJECT OVERVIEW

Wanda Vista Resort Nanchang is a 6-star luxury hotel covering an area of 8.18 hectares, with a total floor area of 45,700 square meters and 203 rooms.

3 南昌万达文华度假酒店夜景
4 南昌万达文华度假酒店总平面图

5

5 南昌万达文华度假酒店鸟瞰图
6 南昌万达文华度假酒店雨棚
7 南昌万达文华度假酒店远景
8 南昌万达文华度假酒店外立面

6

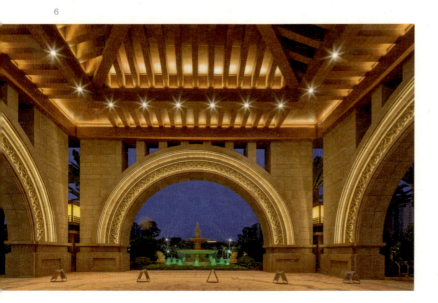

设计理念

南昌万达文华度假酒店采用浪漫清新的地中海风情——以质感墙面、红瓦铺顶的地中海风格作为度假酒店的第一阐述；酒店体量沿湖面充分展开，使建筑获得最大有效沿湖景观，整体简约、大气。营造休闲的生活方式，而自然、简单的生活空间可以让人身心舒畅，感到宁静和安逸，形成九龙湖畔又一城市亮点。

DESIGN CONCEPT

The hotel is designed in a romantic, fresh Mediterranean style, with textured walls, and red tile roofing as the first interpretation of a resort hotel; it is fully expanded along the lake to obtain the maximum efficient landscape along the lake, providing a casual lifestyle. As a natural, simple living space, it makes people feel comfortable, tranquil and cozy, and becomes a new bright city attraction beside the Jiulong Lake.

PART B WANDA CITY 万达城

7

8

生态景观

文华度假酒店景观绿化面积达7.04万平方米,种有四千余棵乔木。酒店掩映在植物之中,主景观以西班牙轴线式花园为主题,采用经典的设计手法及元素以提升景观价值。在中轴线上布置了百米长阶梯跌水和长水渠、台地式户外花园餐吧、趣味性迷宫花园以及四个感官精致的小花园。宾客穿过大堂至酒店后场,可欣赏到气派壮观的中央主水景、中轴长条水景及具有"触觉、味觉、听觉、嗅觉"的体验式西班牙小花园。

ECOLOGICAL LANDSCAPE

Wanda Vista Resort Nanchang has a landscape green area of 70,400 square meters with over four thousand trees. Hidden amidst the greenery, the hotel is designed with a theme of Spanish axial garden, with classic design language and elements adopted to enhance the landscape values. On the axis, there are 100m-long stepped overlapping water and long canals, as well as terrace garden dining bar, maze garden and four exquisite small gardens. Pass through the lobby and enters the hotel's back field, guests can appreciate the magnificent central waterscape, belt-shaped waterscape on the central axis and experiential Spanish small gardens featuring "touch, taste, hearing and smell".

9 南昌万达文华度假酒店水景雕塑
10 南昌万达文华假酒店水景
11 南昌万达文华假酒店景观轴
12 南昌万达文华假酒店景观小品
13 南昌万达文华假酒店喷泉

WANDA REALM RESORT NANCHANG AND PULLMAN NANCHANG WANDA
南昌万达嘉华度假酒店 & 南昌万达铂尔曼酒店

OPENED ON: 28th MAY, 2016
LOCATION: WANDA CITY NANCHANG, NANCHANG
LAND AREA: 9.09 HECTARES
FLOOR AREA: 38,200 SQUARE METERS (WANDA REALM RESORT NANCHANG)
38,000 SQUARE METERS (PULLMAN NANCHANG WANDA)
LANDSCAPE AREA: 79,800 SQUARE METERS

开业时间: 2016 / 05 / 28
开业地点: 南昌市 / 南昌万达城
占地面积: 9.09公顷
建筑面积: 3.82万平方米（南昌万达嘉华度假酒店）
3.80万平方米（南昌万达铂尔曼酒店）
景观面积: 7.98万平方米

14 南昌万达嘉华度假酒店&南昌万达铂尔曼酒店总平面图
15 南昌万达铂尔曼酒店外立面
16 南昌万达嘉华度假酒店鸟瞰图
17 南昌万达嘉华度假酒店立面图

14

15

酒店概况

南昌万达嘉华度假酒店和南昌万达铂尔曼酒店均为五星级酒店，合计占地9.09公顷，建筑面积分别为3.82万平方米及3.80万平方米，拥有客房396间及403间。

PROJECT OVERVIEW

Wanda Realm Resort Nanchang and Pullman Nanchang Wanda are both 5-star hotels, covering a total area of 9.09 hectares, with floor areas of 38,200 square meters and 38,000 square meters, 396 and 403 keys respectively.

PART B　WANDA CITY
万达城

18

18 南昌万达嘉华度假酒店外立面
19 南昌万达嘉华度假酒店建筑景观亭
20 南昌万达铂尔曼酒店外立面

设计理念

酒店采用新古典建筑风格，顺应湖岸走势，强调多重轴线的相互关系，以舒展、流畅的建筑形体关系和走势形成休闲、多样的建筑环境和城市风貌，与酒店风格自然呼应统一。酒店根据自身的用地条件布置，以单排客房为原则，以充分利用湖景资源，保证客房的景观品质，提升酒店价值。建筑立面以层层跌落的坡屋顶作为建筑形态的主要特征，采用经典"三段式"处理手法——底座以小尺度的建筑体块插入，构造宜人的建筑尺度；中段以凸窗台、情境阳台等元素表达度假酒店的特殊品质；顶部高低错落屋面给城市天际线带来丰富的变化。

DESIGN CONCEPT

The hotel, in the neoclassical architectural style, is built to suit the topography at the lakeside. By stressing the relationship between multiple axes, the stretching, fluent architectural and morphological compositions form a relaxing, diversified architectural environment, which echoes in consistency with the urban landscape and hotel style. Based on the hotel's land conditions, guest rooms are arranged in single-rows, so as to make full use of the lake scenery resources, ensure the landscape quality of rooms and enhance property value of the hotel. Facades of the buildings mainly feature the cascading sloping roofs, with the classic "three-section" approach processing technique adopted - on the base small scale of the building blocks are inserted to form appropriate building scale; in the middle section, bay windows, situational balconies and other elements are used to present the unique quality of the resort hotel; on the top, roofs of irregular heights make the city skyline rich and variable.

19

21 南昌万达嘉华度假酒店水景
22 南昌万达嘉华度假酒店景观亭
23 南昌万达嘉华度假酒店水景雕塑
24 南昌万达铂尔曼酒店喷泉
25 南昌万达铂尔曼酒店绿化

生态景观

酒店掩映在绿化面积达6万平方米、乔木两千余棵的景观之中,采取家庭式的度假格调。设计赋予两座酒店不同的空间氛围——万达嘉华度假酒店现代简约,配以干净开阔的大草坪及自然有机的生态花园。万达铂尔曼酒店是新古典的风格,配以几何图案构图的草坪及水法喷泉、故事情节的雕塑;整体被打理精致的自然花溪环绕。

ECOLOGICAL LANDSCAPE

The hotels nestle in the landscape with a green area of 60,000 square meters and more than two thousand trees, providing a family-style vacationing atmosphere. The two hotels are designed with different spatial atmosphere. Wanda Realm Resort Nanchang is in a contemporary minimalist style with clean, open, big lawn and natural, organic ecological garden; while Pullman Nanchang Wanda is in a neo-classical style with geometric-pattern lawns, water fountains as well as sculptures from stories; the entire hotel is embraced by the well-managed naturally by Flowers and Brooks.

26 南昌万达铂尔曼酒店夜景
27 南昌万达嘉华度假酒店夜景
28 南昌万达嘉华度假酒店夜景

夜景照明

酒店的夜景照明效果注重与毗邻的万达文华酒店呼应。照明首先完整表现了建筑的顶部结构和建筑轮廓，实现了酒店群整体的一致性效果。重点照亮了欧陆特色的中层立柱，为酒店立面提供了基本夜景效果。酒店雨篷吊顶造型别致，为通透的透光造型吊顶，既为地面提供足够的功能照明，同时增加雨篷吊顶夜景效果的观赏性，营造了欢乐的度假氛围。

NIGHTSCAPE LIGHTING

The hotels' nightscape lighting effect is designed to cohering with the adjacent Wanda Vista Nanchang. Firstly the lighting perfectly expresses the building's top structure and building contour so as to maintain the consistency of the entire complex. Accent lighting is applied to the middle column with European continental features, bringing up the basic nightscape effect of the hotels' facade. The hotels have specially-shaped awning ceilings, which are transparent to provide sufficient functional lighting for the ground and also increase the visual enjoyability of the ceiling in nightscape, creating joyful holiday experience.

NOVOTEL NANCHANG WANDA AND MERCURE NANCHANG WANDA
南昌万达诺富特酒店 & 南昌万达美居酒店

OPENED ON: 28th MAY, 2016
LOCATION: WANDA CITY NANCHANG, NANCHANG
LAND AREA: 4.69 HECTARES
FLOOR AREA: 32,400 SQUARE METERS (NOVOTEL NANCHANG WANDA)
32,400 SQUARE METERS (MERCURE NANCHANG WANDA)
LANDSCAPE AREA: 41,000 SQUARE METERS

开业时间： 2016 / 05 / 28
开业地点： 南昌市 / 南昌万达城
占地面积： 4.69公顷
建筑面积： 3.24万平方米（南昌万达诺富特酒店）
3.24万平方米（南昌万达美居酒店）
景观面积： 4.1万平方米

29

29 南昌万达诺富特酒店&南昌万达美居酒店总平面图
30 南昌万达诺富特酒店&南昌万达美居酒店鸟瞰图
31 南昌万达诺富特酒店&南昌万达美居酒店外立面

酒店概况

南昌万达诺富特酒店和南昌万达美居酒店均为四星酒店，占地面积4.69公顷，建筑面积均为3.24万平方米，分别拥有房间480间和504间。

PROJECT OVERVIEW

Both Novotel Nanchang Wanda and Mercure Nanchang Wanda are 4-star hotels, covering a total area of 4.69 hectares, with floor area both being 32,400 square meters, 480 and 504 keys respectively.

30

设计理念

酒店采用北美风格，位于滨湖酒吧街西侧，拥有大窗、阁楼、坡屋顶及丰富的色彩和流畅的线条。立面为"三段式"——底部立面采用石材，配合开敞的大窗，形成明亮时尚的现代感；主入口为Craftman式雨篷，轻松时尚。建筑中部选用米黄色涂料，色调温暖舒适；墙面采用Stucco设计手法，将大面积墙面打破，配合宽大的高窗，有利于欣赏窗外美景。顶部木质感材料与蓝色屋顶配合柔和，营造轻松自然的酒店氛围。

DESIGN CONCEPT

Sitting on the west side of the Bar Street at lakefront, the Hotel is designed with a North-American style, with large windows, attics, sloping roof and rich color as well as smooth lines. The "tripartite composition" approach is adopted in facade design, where stones are used for the bottom facade in combination with open large windows, forming bright, stylish modern feelings; casual, fashionable Craftman style awning is arranged at the main entrance. In the middle section of the building warm, comfortable beige color paint is applied; Stucco design technique is adopted for the wall surface, so that the large area of the wall surface is broken up by large, wide windows, facilitating viewing sights outside. On the top woody materials are used on the blue roof, creating a relaxing, natural atmosphere for the hotel.

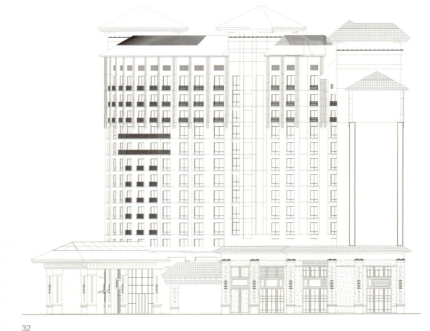

34

生态景观

酒店景观绿化面积达4.1万平方米，配以上千棵乔木，将酒店掩映在植物之中。入口的景观采用现代雕塑水景呼应美式草原建筑风格；后花园则以借景方式让宾客置身湖滨公园之中。

ECOLOGICAL LANDSCAPE

The Hotel has a landscape green area of 41,000 square meters with over one thousand trees, covering the hotel's architecture with their shades. Landscape at the entrance is made of modern sculptures in waterscape as an echo of the American-style prairie architectural style; while the back garden presents a rich lakeside park atmosphere via view-borrowing.

32 南昌万达诺富特酒店&南昌万达美居酒店总立面图
33 南昌万达诺富特酒店&南昌万达美居酒店绿化
34 南昌万达美居酒店绿化
35 南昌万达美居酒店绿化

35

05

WANDA CITY NANCHANG - BAR STREET
南昌万达城酒吧街

LOCATION : WANDA CITY NANCHANG, NANCHANG
LAND AREA : 3.49 HECTARES
FLOOR AREA : 20,000 SQUARE METERS
LANDSCAPE AREA : 14,000 SQUARE METERS
PLANNED FUNCTIONS : BAR, RELAXATION, ENTERTAINMENT

开业地点： 南昌市 / 南昌万达城
占地面积： 3.49公顷
建筑面积： 2.0万平方米
景观面积： 1.4万平方米
规划功能： 酒吧，休闲，娱乐

酒吧街概况

南昌万达酒吧街沿九龙湖一字展开，共有酒吧街1号～酒吧街5号共5栋单体，占地3.49公顷，建筑面积2.0万平方米，共有18个酒吧及两座酒楼。

PROJECT OVERVIEW

The Bar Street of Wanda City Nanchang runs along the Jiulong Lake, with altogether 5 individual buildings, namely Bar Street No.1 to Bar Street No. 5. It covers an area of 3.49 hectares and a floor area of 20,000 square meters, with a total of 18 bars and two restaurants.

1 南昌万达城酒吧街总平面图
2 南昌万达城酒吧街立面图
3 南昌万达城酒吧街远景
4 南昌万达城酒吧街外立面
5 南昌万达城酒吧街建筑立面图
6 南昌万达城酒吧街外立面

设计理念

南昌万达城酒吧街沿九龙湖畔排开，入口广场沿路做出景观退让。建筑体块简洁而富有内涵，为地中海风格，与周边酒店群建筑风格相呼应；立面材料搭配丰富，细部装饰清新、恬淡。酒吧街将自然景观与消费业态完美结合，并借九龙湖水系使得每间酒吧都能得到宽阔的视野。

DESIGN CONCEPT

The Bar Street of Wanda City Nanchang runs along the shore of the Jiulong Lake, the entrance square is recessed along the roadside for landscape. The building blocks, featuring simplicity and rich connotations, adopts the Mediterranean style that is consistent with the architectural style of the surrounding hotel complex; facades are made up of various types of materials arranged together with fresh, tranquil detailed decorations. The Bar Street is a perfect integration of natural landscape and building function. Every bar enjoys a wide view by virtue of the water system of the Jiulong Lake.

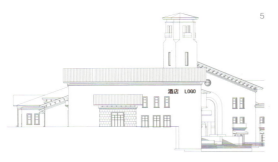

7

生态景观

南昌万达城酒吧街拥有景观绿化面积约1.4万平方米，辅以三千余棵乔木。临水的平台提供观览湖景的适宜环境，充满西班牙气息的街道家具和地面铺装又平添许多异国情调。

ECOLOGICAL LANDSCAPE

The Bar Street has a landscape green area of about 14,000 square meters with over three thousand trees. The waterfront platform offers a suitable environment for viewing the lake sceneries; Spanish-style street furniture and pavement highlight the exotic atmosphere to a large extent.

7 南昌万达城酒吧街景观雕塑
8 南昌万达城酒吧街建筑外立面

8

PART B WANDA CITY
万达城

06

LAKESIDE GREEN BELT OF THE JIULONG LAKE, WANDA CITY NANCHANG
万达南昌九龙湖滨湖绿带

OPENED ON: SEPTEMBER 2016
LOCATION: WANDA CITY NANCHANG, NANCHANG
LAND AREA: 36.64 HECTARES

开业时间：2016 / 09
开业地点：南昌市 / 南昌万达城
占地面积：36.64 公顷

1 万达南昌九龙湖滨湖绿带总平面图
2 万达南昌九龙湖滨湖绿带景观

规划概述

景区的沿湖步道景色以自然生态的手法来营造，表现南昌不同的文化及自然特色，从三星酒店北侧市民公园沿湖串联至六星酒店北侧私家码头岛，沿途各景点风景各异，有各自代表的场所气氛，色彩花色美不胜收，也使得九龙湖沿岸步道有多样的节点形态。

PLANNING OVERVIEW

The scenes along the footpath surrounding the lake is created in a natural ecological approach, so as to present the different cultural and natural features of Nanchang; from the public park on the north side of the 3-star hotel to the private wharf island on the north side of the 6-star hotel, there are distinct beautiful sceneries showing different atmosphere of their respective spaces, enriching the lakeside trail with varied node forms.

WANDA CITY HEFEI
合肥万达城

|01

MASTER PLAN OF WANDA CITY HEFEI
合肥万达城总体规划

PLANNING DURATION: MARCH 2013 TO SEPTEMBER 2015
LOCATION: HEFEI, ANHUI PROVINCE
PLANNED AREA: 157 HECTARES
GROSS FLOOR AREA: 3.47 MILLION SQUARE METERS
FUNCTIONS: CULTURAL PROJECTS: MOVIE PARK, MOVIE CITY
　　　　　　　TOURISM PROJECTS: OUTDOOR THEME PARK, INDOOR WATER PARK
　　　　　　　HOTEL PROJECT: HOTEL COMPLEX
　　　　　　　COMMERCIAL PROJECTS: LARGE SHOPPING CENTER, BAR STREET, OUTDOOR COMMERCIAL STREET, STREET SHOPS

规划时间： 2013 / 03 – 2015 / 09
规划位置： 安徽省 / 合肥市
规划面积： 157 公顷
建筑面积： 347 万平方米
规划功能： 文化项目：电影乐园、影城
　　　　　旅游项目：室外主题乐园、室内水乐园
　　　　　酒店项目：酒店群
　　　　　商业项目：大型购物中心、酒吧街、室外商业街、沿街商铺

一、缘起

合肥市正成为"长三角经济圈"的新动力。合肥市滨湖新区是合肥市突破现有城市格局、空间结构，由"单中心模式"向"多中心模式"发展的重要突破口；大量行政、金融、商业、居住等功能纷纷迁入；同时积极引入外部资源，打造生态型、高端化的尖端产业聚集区。万达集团顺势而为，将合肥的区位优势和万达强大的产业创新实力相结合，创造"长三角经济圈"的第一个"万达城"。

二、万达城——国际一流文化旅游度假区

万达拥有独创的世界一流的文化旅游建设经验，结合当地特色，融入最新科技，为合肥市量身打造全球顶级的文旅新城。"万达城"借助巢湖的生态优势，坐拥"八百里巢湖风光"以及位于开发区"黄金地带"的先机，将进一步激活合肥城市活力，成为合肥"创新驱动新引擎"和"城市发展新名片"。

以"合肥万达城"为代表的"升级版万达城"，业态更齐全，设备更先进，规划布局更集中；业态包括"四大板块"：文化项目——电影乐园、影城；旅游项目——室外主题乐园、室内水乐园；酒店项目——万达文华酒店（6星）、万达嘉华度假酒店（5星）、万达铂尔曼酒店（5星）、万达美居酒店（4星）和万达诺富特酒店（4星）；商业项目——大型购物中心、酒吧街、室外商业街、沿街商铺。

新一代"万达城"在宏观规划层面进行业态的有机整合和聚集，最大化发挥"万达"潜力，聚合城市旅游资源，提高产业发展层次，催生和促进多种现代服务业态的发展繁荣。新一代"万达城"还深入挖掘文化要素，与当地文化结合更加紧密，推动文化和旅游深度融合。新一代"万达城"旅游体验感进一步加强，突出独特性和唯一性，为城市量身打造不可复制的特色高端旅游度假区，构建完善的文化旅游产业链，引领旅游度假新潮流。

I. ORIGIN

Hefei is becoming a new driving force in the "Yangtze River Delta Economic Zone". Hefei Binhu New District is an important breakthrough for the city to break the existing urban pattern and spatial structure, and to develop from the mode of one center to the mode of several centers; a large number of administrative, financial, commercial, residential and

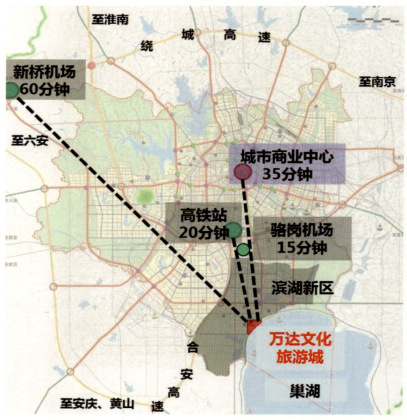

图1 合肥万达城区位图

图2 合肥万达城总平面图

三、合肥万达城——"交通优越、集约布局、生态新城、文化为魂"的"四位一体"

合肥万达城位于合肥市滨湖新区，南邻巢湖，距合肥城市商业中心35分钟车程，距离高铁站20分钟车程，距骆岗机场、新桥机场分别为15分钟和60分钟车程（图1）。于2013年正式开工，2016年9月全面建成开业，占地面积157公顷；其中文化旅游区占地92公顷，建筑面积100万平方米，可同时容纳5万名游客，日接待量10万人次，年接待游客2000万人次（图2~图5）。

1. 交通优越——外部通达，内部有序

外部方面——合肥是华东地区交通最便利的节点城市之一，武汉、南京都在两小时辐射圈内；借助"长三角经济圈"的优势，使合肥成为区域性国际航空港。通过环湖北路和地铁一号线，"万达城"可便捷到达合肥市主要功能区。从机场、高铁、长途汽车总站，均可乘坐地铁1号线直达"万达城"。

内部方面——"万达城"内部交通组织有序，万达茂南北两个入口广场与地铁站无缝对接，吸纳了大量人流。万达茂和室外主题乐园共用出入口，不仅节约土地，也

other functions will move in; at the same time it will actively attract external resources to build an eco-friendly, high-end cluster with top industries. Taking advantages of the situation, Wanda Group combines the geographical advantages of Hefei with its strong industrial innovation strength to build the 1st "Wanda City" at the "Yangtze River Delta Economic Zone".

II. WANDA CITY - A FIRST-CLASS INTERNATIONAL CULTURAL TOURISM RESORT

Wanda has unique experience of constructing world-class cultural tourism projects. When combined with local features and the latest technologies, it will build a top of the world cultural and tourism city tailored for Hefei. By taking advantages of Chaohu's ecological environment, the superb location of the "Eight-hundred Mile Lake-scape", and the "golden place", "Wanda City" will further activate the vitality of Hefei, and will become "New Engine" and "New Business Card of Urban Development".

The "upgraded Wanda city" represented by "Hefei Wanda City" will have more business formats, more advanced equipments and a more concentrated layout. The formats include "Four Major Items": cultural projects - Movie Park and movie city; the tourism projects - outdoor theme park and indoor water park; hotel projects – Wanda Vista(6-star), Wanda Realm(5-star), Wanda Pullman(5-star), Wanda Mercure (4-star) and Wanda Novotel(4-star); commercial projects - shopping malls, bar street, outdoor commercial street and street shops.

The new generation of "Wanda City" will properly gather and integrate all business formats at the macro planning level to maximize the potential of "Wanda City", it will aggregate the city's tourism resources, and improve industrial development level and promote the prosperity of a variety of modern services. The new generation of "Wanda City" will further explore cultural elements, and work closely with local culture to promote in-depth infusion of culture and tourism. Tourist experience at the new generation of "Wanda City" will be further improved to highlight its uniqueness, to build featured top-level tourist resorts that are tailored for the city and could never by copied, to build a complete cultural tourism industrial chain, and to lead the new trend for tourism and vocations.

III. WANDA CITY HEFEI - FOUR CHARACTERISTICS OF THE CITY: SUPERB ACCESSIBILITY, INTENSIVE LAYOUT, ECOLOGICAL NEW CITY AND CULTURAL ESSENCE

Wanda City Hefei is located at Hefei Binhu New District, and the Chaohu Lake is at the south of it. It's 35 minutes' drive to the city's commercial center, 20 minutes' drive to the high-speed rail station, 15 minutes' drive to Luogang Airport and 60 minutes' drive to Xinqiao Airport (Fig.1). The construction was commenced in 2013, and completed in September 2016. It occupies 157 hectares of land, in which the culture and tourism area takes 92 hectares, and the floor area is 1 million square meters. It can accommodate 50,000 tourists, and receive 100,000 tourists in one day and 20 million tourists in a year(Fig.2-Fig.5).

图3 合肥万达城东南鸟瞰图

促进了万达茂内部业态和室外主题乐园的衔接。酒店区内部的环形健身步道连接巢湖风光带；健身休闲的同时可以享受精致的酒店景观。

2. 集约布局——功能复合，创新文旅

"合肥万达城"荟萃了文化、旅游、酒店、商业"四大类核心产品"。各业态混合兼容，通过"紧凑的和复合的开发"集约利用土地资源，满足多样化的旅游、休闲、度假、娱乐体验，共同构成充满活力、功能复合的国际旅游度假区。

（1）文化项目

"合肥万达城"倾力打造独特的文化项目。其中室内电影乐园设置"飞越安徽"、3D互动剧场、3D体验剧场等世界最新电影科技娱乐项目，带给观众身临其境的飞翔体验；电影城设有12个厅，2000个座位，是目前国内设施最先进、档次最高的电影城之一；室内演出舞台由全球顶级建筑艺术大师马克·费舍尔先生策划导演，将民族艺术，结合科技的灯光音效、舞蹈、杂技、跳水等多种表演形式完美呈现，将浓郁的中国徽文化与现代科技巧妙融合。

（2）旅游项目

旅游项目是"合肥万达城"的核心功能，包括室外主题乐园和室内主题乐园。室外主题乐园占地面积约40公顷，与万达茂入口广场和地铁站结合布置，最大化提升游客的可达性。国际顶尖主题乐园设计公司FORREC担纲设计的万达乐园，是全球唯一以"徽文化"为主题的乐园。室内水乐园引进全球顶尖科技的龙卷风、高速竞技滑梯、过山车等内容，作为华东地区唯一恒温水乐园，给游客带来全天候的水中欢乐。

（3）酒店项目

度假酒店群占地约30公顷，外滨巢湖，内临人工湖，巢湖美景和人工湖景观尽收眼底。17公顷的人工湖和超过1公里的人工沙滩，创造出"一体化"的"水系—绿茵生态环境"。酒店群以围合式布局为主，通过对建筑形态与景观朝向的匠心经营，呈现出层次丰富、各具特色的庭院及滨水空间，将合肥的旅游和会议接待提升到国际水平。

1. SUPERB ACCESSIBILITY - REACHING THE OUTSIDE IN ALL DIRECTIONS AND ORGANIZING THE INSIDE ORDERLY

Reaching the outside - Hefei is one of the cities in East China that have the most convenient accessibility. Wuhan and Nanjing can be reached in less than two hours. The advantage of "Yangtze River Delta Economic Zone" will make Hefei the international airport for the region. One can conveniently reach any of the functional areas in Hefei from "Wanda City" via the Northern Lake Ring Road and Subway Line 1. People can go to "Wanda City" by taking Subway Line 1 from the airport, high-speed rail station or coach station.

Organizing the inside - The transportation within "Wanda City" is orderly organized. The entrance squares at the south and north sides of Wanda Mall connect to the subway station seamlessly taking a lot of people. Wanda Mall and the outdoor Theme Park share one entrance, which not only saves land but also promotes the connection between the business formats inside Wanda Mall and those in the Theme Park. The circular walkway in the hotel zone is connected to the landscape of Chaohu so that guests can enjoy the exquisite landscape while taking recreation.

2. INTENSIVE LAYOUT - INTEGRATED FUNCTIONS AND INNOVATIVE CULTURAL TOURISM

"Wanda City Hefei" has four core products - culture, tourism, hotel and commerce. All business formats are mixed together and compatible with each other. By means of "compact and integrated development", the land is intensively used to meet diversified requirements of tourism, recreation, vacation, and amusement, composing a vital international tourism resort with integrated functions.

(1) Cultural Projects

"Wanda City Hefei" makes efforts to build unique cultural projects. The world's latest movie technology amusement shows, like "Flying over Anhui", 3D interactive theater, 3D experience theater, are provided in the indoor movie park, bringing true experiences to the audience as if they are flying in the sky. There are 12 screens and 2000 seats in the movie city. It is one of the movie cities in China with the most advanced facilities and highest level of service. Designed by Mark Fisher, the master of architecture, the indoor performance stage can present national arts with lighting, sound effect, dancing, acrobats, diving and other forms of performances, and it smartly integrates the deep Chinese Hui culture with modern technologies.

(2) Tourism Projects

Tourism projects, the core function of "Wanda City Hefei", include the outdoor theme park and indoor theme park. Covering about 40 hectares of land, the outdoor theme park integrates with Wanda Mall entrance square and the subway station to maximize the accessibility. The park is designed by FORREC, the top theme park design company. It is

图4 合肥万达城西南鸟瞰图

图5 合肥万达城西北鸟瞰图

（4）商业项目

"万达城"商业项目包括大型购物中心、酒吧街、室外商业街和沿街商铺"四大板块"。万达茂引进近二百个品牌商家，特别引进50多家不同风味的全球美食餐厅，给游客带来"舌尖上的享受"。商业项目为本地居民和国内外游客提供国际一流的休闲购物体验，是安徽特色鲜明、现代感强烈、品质突出的商业中心。酒吧街依湖而建，建筑面积2万平方米，引进二十余家知名的国内外酒吧、音乐吧品牌。

3. 生态新城——海绵城市、生态渗透

滨湖新区是合肥市重点打造的生态新城。借助巢湖的生态基础，"合肥万达城"内部生态绿地系统与外部道路绿化、巢湖沿岸绿化有机衔接，打造多条生态绿廊、绿化斑块，构成滨湖新区与巢湖生态衔接的通道。"合肥万达城"采用"海绵城市"建设理念，与合肥市共同将主要干道庐州大道打造成全市第一条绿色生态道路和滨湖新区第一条"海绵城市"试点市政道路。通过屋顶绿化滞留雨水的合理利用，起到节能减排、缓解热岛效应的功效。在小区规划中，将有条件的小区绿地"沉下去"，让雨水进入下沉式绿地进行调蓄、下渗与净化，减少直接通过下水道排放。

4. 文化为魂——根植徽派、人文新城

"合肥万达城"在符合现代化新区发展和现代生活方式的同时，完美融入合肥历史文化的精髓。合肥有两千多年的历史，素以"三国故地、包拯家乡"而闻名海内外，一直是江淮地区重要的行政中心和军事重镇。"合肥万达城"展现了"包公文化"、"三国文化"、"药王文化"、"宋词文化"等为代表的徽派"八大文化"。

"合肥万达城"各个业态结合"徽文化"，主题乐园将徽派戏曲、历史故事融入游乐项目，集中安徽各地著名老字号、民俗商品、特色餐饮，成为全球首座大型徽文化主题乐园；"万达茂"购物建筑设计寓意徽州自古为诗书礼乐之邦、宣纸发明之地；新徽派建筑风格的度假酒店群将徽州传统文化与人文景观复原，水系景观与亭台楼阁巧妙结合，处处充满徽派神韵。

四、结语

"合肥万达城"借势安徽省政府迁入滨湖新区的发展契机，积极融入新区整体功能定位，以万达先进的文化旅游业态引领滨湖新区板块创新发展，借助万达品牌形成合肥创新产业高地。"合肥万达城"成为活力四射的"时尚新城"和"合肥文化旅游地标，"助力合肥成为中国文化旅游名城和世界级旅游目的地。

the only theme park in the world that features "Hui Culture". Cyclone, hi-speed slide, water magnetic roller and other facilities backed with top technologies in the world are introduced in the indoor water park. As the only thermostatic water park in East China, it will bring all-weather fun to the guests.

(3) Hotel Projects

The hotels cover about 30 hectares. Next to Chaohu Lake and the artificial lake, they are the best choice for enjoying the landscape. 17 hectares of artificial lake and over 1 km of artificial sand beach offer an integrated "water-green environment". The hotel complex is mainly in an enclosed layout. By managing architectural form, landscape and orientation, it shows courtyards and waterfronts with rich gradations and features, which would improve the tourism, conference and reception events in Hefei to international level.

(4) Commercial Projects

The commercial projects of "Wanda City" include four categories: large shopping center, bar street, outdoor commercial street and street shops. Wanda Mall has introduced nearly 200 brands and more than 50 restaurants with a variety of delicious food from the world to bring "Pleasure to the Tongue" to the guests. The commercial projects will offer recreation and shopping experience at top international level to local residents, as well as domestic and international tourists. It will become a featured, contemporary commercial center with high quality. The bar street is around the lake with 20,000 square meters of floor area. More than twenty famous brands of domestic and international bars and music bars will be introduced to open business here.

3. NEW ECOLOGIC CITY - SPONGE CITY AND ECO-PENETRATION

Binhu New District is a new ecologic city built by Hefei government with efforts. Borrowing from the ecologic foundation of Chaohu, the ecologic greenbelt system in "Wanda City Hefei" will be connected to external roads and greenbelts and the landscape around Chaohu, building several ecologic green corridors and areas and forming an ecologic passage connecting Binhu New District and Chaohu. "Wanda City Hefei" adopts the concept of "sponge city", and will work with Hefei government to build the main road Luzhou Road into the first green, ecologic road in the city and the first "sponge city" pilot municipal road in Binhu New District. Rational utilization of the rainwater retained at the rooftops and greenbelts will save energy, reduce emission and relieve heat island effect. When the community is designed, if the conditions permit the greenbelts in the community will be lower than the road surface so that rainwater can enter the area for regulation, storage, infiltration and purification, which can reduce the direct discharge from the sewer.

4. CULTURAL ESSENCE - ROOTED ON HUI CULTURE AND BUILD A NEW CULTURAL CITY

While conforming to the development of the modern new district and modern life style, "Wanda City Hefei" has perfectly integrated the essence of the history and culture of Hefei. Having more than two thousand years of history, Hefei is famous at home and abroad for its importance in the Three Kingdom period and the hometown of Bao Zheng. It has been an important administrative center and military fortress at Jianghuai Region. "Wanda City Hefei" has manifested "eight cultural elements" of Hui Style, including "Bao Gong Culture", "Three Kingdom Culture", "King-of-Medicine Culture", "Poem-at-Song-Dynasty Culture".

The business formats at "Wanda City Hefei" have combined a lot of "Hui Culture". The Theme Park integrates traditional Chinese opera and historic stories of Hui Style into the entertainment projects, and it has included famous old brands, folk commodities and featured food, and becomes the first large-scale theme park with Hui Culture in the world. The architectural design of "Wanda Mall" highlights that Huizhou has been a town of learning and civilization, and the place where Xuan Paper is invented. The hotel complex in the architectural style of new Hui Style restores the traditional culture and cultural landscape of Huizhou, and smartly combines water landscape, pavilions and buildings, showing the essence of Hui Style everywhere.

IV. CONCLUSION

By taking the opportunity that Anhui Provincial Government moves into Binhu New District, "Wanda City Hefei" actively integrates into the overall functions of the new district, taking the lead in the innovation and development of Binhu New District with Wanda's advanced cultural and tourism business, and raising a new industrial high ground in Hefei by using Wanda brand. "Wanda City Hefei" will become an energetic "vogue city" and "landmark in Hefei for culture and tourism", and helps Hefei to become a famous cultural and touristic city in China and a touristic destination in the world.

02

WANDA MALL HEFEI
合肥万达茂

OPENED ON: 24th SEPTEMBER, 2016
LOCATION: BINHU NEW DISTRICT, HEFEI
LAND AREA: 20.8 HECTARES
FLOOR AREA: 187,600 SQUARE METERS

开业时间：2016 / 09 / 24
开业地点：合肥市 / 滨湖新区
占地面积：20.8 公顷
建筑面积：18.76 万平方米

1 合肥万达茂区域图
2 合肥万达茂总平面图
3 合肥万达茂外立面

万达茂概述

合肥万达茂位于合肥市新城开发区滨湖核心区域——滨湖新区庐州大道与南宁路交口，西靠井冈山路，西侧、东侧和南侧均为市政道路——人口密集、交通便利，城市发展区位优势明显，设有地铁一号线"万达城站"。万达茂是由室内步行街、娱乐楼、电影乐园和水乐园组成的大型娱乐性综合体项目；其中室内步行街建筑面积13.93万平方米（含配套1.42万平方米）、电影乐园1.63万平方米，水乐园3.20万平方米。合肥万达茂，将构建成合肥的文化、休闲、购物和娱乐交流中心。

PROJECT OVERVIEW

Wanda Mall Hefei is located at the core of Binhu District in the New Town Development Zone of Hefei, which is at the intersection of Luzhou Avenue and Nanning Road in the Binhu District. Next to Jinggangshan Road in the west, with municipal roads at its west, east and south sides, Wanda Mall is a densely populated area with convenient transportation and advantageous location in the city. Wanda Mall is a large-scale entertainment complex consisting of the indoor pedestrian streets entertainment building, Movie Park and Water Park; among them the indoor pedestrian street covers a floor area of 139,300 square meters (including 14,200 square meters of facilities), the Movie Park 16,300 square meters and water park 32,000 square meters respectively. Wanda Mall Hefei will become the cultural, recreational, shopping and entertainment center of Hefei.

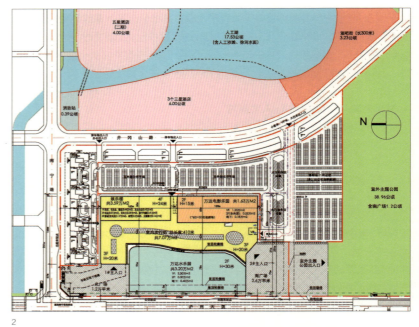

设计管控

合肥万达茂外观设计立意取自"书卷"——片片"书页"层层堆叠——屋面强调层次,立面展现动势,入口则是节奏的高潮。为了强化万达茂入口的视觉识别性,让宾客"第一眼"就被其浓郁的人文气息所感染,设计团队将怀素狂草体"合肥文化八景"、"安徽文化十景"书法作品通过打点玻璃工艺加以运用,实现中国传统文化与现代工艺的碰撞,产生了别样夺目的效果。实体样板完成后,设计团队仔细推敲,将文字效果、印章位置等处进一步完善,达到"密不透风,疏可跑马"的艺术效果;衬墙为体现安徽"山水壮阔"的意向,改用竖向细条构图,并在细条后部隐藏灯带,夜晚变幻出山水画面效果。完成后的合肥万达茂入口舒展流畅,富有律动、一气呵成,达到了预期效果。

DESIGN MANAGEMENT

The design concept of Wanda Mall Hefei comes from "book volume" - pages of the "book" are stacked layer by layer - the rooftop emphasizes layering, the facade presents momentum, and the entrance marks the peak of the rhythm. In order to make the entrance of Wanda Mall more recognizable, so that the visitors will feel its profound cultural atmosphere "at the first sight", the design team incorporated Huaisu's cursive calligraphy works "Eight cultural views of Hefei" and "Ten cultural views of Anhui" into the design utilizing the glass dotting technique, so as to produce a super brilliant effect via conflicts between Chinese traditional culture and modern technologies. After the material sample was completed, the design team scrutinized it deliberately to further improve the text effect and stamp position, so as to create an artistic effect that is "when tightened airless, when loosened a hourse can run through"; vertical thin strip patterns were adopted on substrate walls to reflect "majestic landscapes" imagery of Anhui, with light bars hidden behind the thin strips, presenting landscape images at night. The entrance of Wanda Mall was in a smoothly stretching rhythm, which was exactly the expected effect.

4

5

万达茂外装

合肥万达茂建筑外立面取意于"徽州书韵,山水长卷"——造型层层叠叠,寓意翻开的片片"书页";在层层堆叠的基础上,用灯带勾勒出数道从屋面一直延伸至立面的大弧线,既与屋面"书卷"整体动势相呼应,又加强了"书卷"的层次感——体现对"徽州文化"的凝练、对"书卷理念"的坚持;精雕细琢的"匠人工法",将徽州"山与水"巧妙地融入其中,使得现代建筑技艺与传统文化取得和谐与统一的效果。由徽州当地名家创作的《合肥八景》《安徽十景》,通过图案数字化技术和现代工艺,在万达茂南北主入口立面上得到解构与重组,展现了"徽派文化"精髓,又不失现代美学"构成美"的神韵,真正做到了对经典的创新演绎。

FACADE DESIGN

The design concept for the facade of Wanda Mall Hefei came from "Book charms of Huizhou, long scrolls of landscapes" - the design is comprised of layers over layers, which mean the open "pages" of a book; in addition, light bars are used to form many long arcs extending from the roof toward the facade, which echo the overall trend of the "volume" on roof and also enhance the stratification effect of the "volume" - presenting extraction of "Huizhou culture" as well as the persistence on "book concept"; via the refined "craftmanship", the "mountains and waters" of Huizhou were incorporated in the design in a skillful way. "Eight Views of Hefei" and "Ten scenes of Anhui" created by Huizhou local artists, are deconstructed and reorganized in the facades of the main entrance of Wanda Mall, by using pattern digitalization technology and modern techniques. It shows the essence of "Hui Culture" yet maintains the "beauty of composition" in modern aesthetics at the same time. It interprets classics in an innovative way.

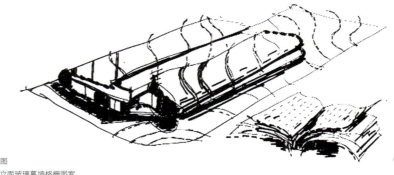

4 合肥万达茂主入口
5 合肥万达茂外立面
6 合肥万达茂概念草图
7 合肥万达茂主入口立面玻璃幕墙格栅图案
8 合肥万达茂主入口雨棚

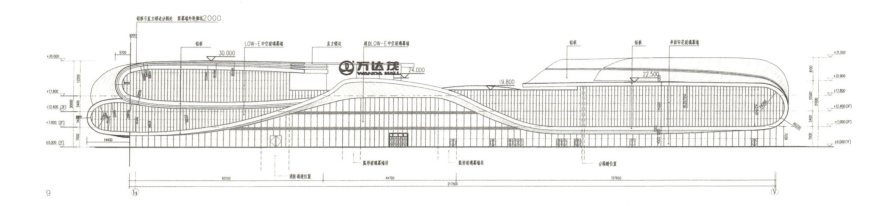

PART B WANDA CITY 万达城

9 合肥万达茂立面图
10 合肥万达茂外立面
11 合肥万达茂外立面

万达茂内装

合肥万达茂是汇集了室内水乐园、电影乐园、商业和影院等多种业态于一体的大型娱乐性综合体项目。安徽以盛产宣纸、徽墨和毛笔而闻名，因此将中国绘画中的"写意山水"融合在室内设计中。以"毛笔在宣纸上游走的印记"作为设计主题，将整个室内空间作为一个画卷，空间的造型处处表现出"中国画的笔迹"。石材地面拼花成为一幅巨大的画卷——两入口之间，以连续不断的笔迹完成抽象水墨画面，画卷的笔触聚散于两个中庭。其内容为徽州山水景色，白色石材作为"画卷"的底色，图案为黑色、深灰、浅灰的地面石材绘就。立面设计同样如宣纸上笔墨丹青，描绘出一幅巨型的山水画。其中椭圆中庭起伏的造型，在光线变化之中把山水画的墨迹演化得出神入化。而圆中庭是镂空的山水图案，通过背后的灯光展现出波光粼粼的巢湖，也映衬了"徽州八景"之一的"巢湖夜月"。

INTERIOR DESIGN

Wanda Mall Hefei is a large-scale entertainment complex consisting of multiple business types including indoor water park, movie park, shopping mall and cinema. Anhui has been known for Xuan Paper, Huizhou Inksticks and Writing Brushes. Therefore, "impressionistic landscape" in Chinese paintings became an element in interior design. With "marks of writing brush wandering on the Xuan Paper" as the theme of design, the entire interior space is just like a picture scroll, in which spatial shapes were planned to show "handwriting of Chinese paintings" everywhere. Patterns on the stone floor make up a huge picture scroll, which is right between the two entrances; with continuous strokes mainly converging in the two atriums an abstract ink-wash painting was presented. The painting is about Huizhou landscape sceneries, with white stone as the underlying painting of the picture, and patterns made of black, dark gray and light gray ground stones. The facade design, also like ink-wash paintings on Xuan Paper, is a huge drawing of landscapes. Among them, the undulating contours shown in the oval atrium, vividly simulate the ink marks of landscape paintings under the changing lights. In the round atrium, a hollowed-out landscape pattern is presented; against the changing lights in the background, a sparkling Chaohu Lake unveils, perfectly setting off "Night Moon at Chaohu", one of the "Eight Sceneries of Huizhou".

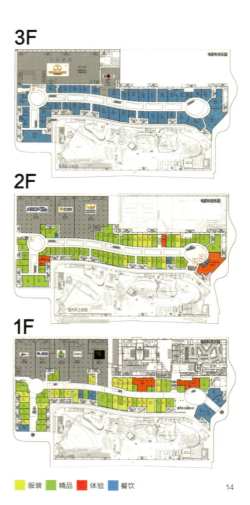

12 合肥万达茂室内步行街
13 合肥万达茂室内步行街电梯
14 合肥万达茂商业落位图
15 合肥万达茂室内步行街中庭

16

17

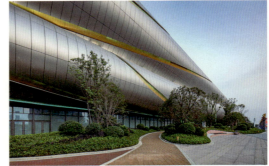
18

万达茂景观

合肥万达茂景观面积（包括建筑屋面美化）为18.49公顷，分为主广场区、带状广场区、后勤区及地面停车场区。景观设计把建筑"书卷"的整体概念进行延伸，将文化符号融入景观中进行呈现——场地铺装以"山水"为肌理，并融入"徽剧脸谱"的元素；南北两端广场上分别以"书香卷轴"与"徽州山水"为主景雕塑形成标志物；北广场水景雕塑提取"千古徽茶"造型；南广场花坛融入"皖印"主题雕塑，水池则融入"文房四宝"中墨砚的造型；带状广场景观雕塑结合"荷香鹤舞"元素。

LANDSCAPE DESIGN

wanda mall Hefei covers a landscape area of 18.49 hectares, which is divided into the primary square, stripe-shaped square, service area and surface parking area. Landscape design extends the overall concept of "book volumes", and renders the cultural symbols by integrating them into the landscape - with "landscape" as the texture for pavement, and combining the "Anhui opera facial makeup" element; on the squares at the north and south ends, main feature sculptures of "scholarly scroll" and "landscape" are erected respectively as the landmark; on the northern square there is waterscape sculpture "eternal Anhui tea"; on the southern square the themed sculpture "Anhui imprinting" is incorporated in the flower beds, and the shape of "ink and inkstone", two of "the scholar's four jewels" can be seen in the pool; the landscape sculpture on the stripe-shaped square adopts the "lotus fragrance and crane dancing" element.

19

16 合肥万达茂地面铺装
17 合肥万达茂导视牌
18 合肥万达茂绿化
19 合肥万达茂景观雕塑
20 合肥万达茂鸟瞰夜景
21 合肥万达茂主入口夜景

万达茂夜景

合肥万达茂夜景照明的创意设计主题为"尺幅千里，墨韵留香"，通过多元化的设计手法及"点"、"线"、"面"结合的灯光设计方式，不仅将水墨书画、印章等"徽文化"融入建筑形态中，也将山峦层叠的意境呈现，共同组成一幅完整兼具山水意境的画卷。南、北两个入口立面是设计的"画龙点睛"之处。万达茂夜景灯光，伴随"书卷造型"、携"徽文化"建筑的古韵魅力尽显风采。

NIGHTSCAPE DESIGN

The nightscape lighting design of Wanda Mall Hefei takes the theme of "in a scale of one thousand miles, the lingering charms of ink can be felt"; through various design techniques and lighting design the combines "points", "lines" and "surfaces", not only Chinese ink painting and calligraphy, stamps and other "Anhui cultures" are added into the architectural form, but also a complete, artistic picture scroll is shown in front of us, rendering the layered mountains. Facades of the south and north entrances are the "finishing touch" of the design. In the nightscape lights of Wanda Mall, the antique charms of the building, together with "book volume shape" and "Anhui cultures" are perfectly displayed.

03

HOTEL COMPLEX OF WANDA CITY HEFEI

合肥万达城酒店群

OPENED ON: 24th SEPTEMBER, 2016
LOCATION: BINHU DISTRICT, HEFEI, ANHUI PROVINCE
LAND AREA: 22.5 HECTARES
FLOOR AREA: 216,000 SQUARE METERS
LANDSCAPE AREA: 171,800 SQUARE METERS

开业时间: 2016 / 09 / 24
开业地点: 安徽省 / 合肥市滨湖区
占地面积: 22.5 公顷
建筑面积: 21.6 万平方米
景观面积: 17.18 万平方米

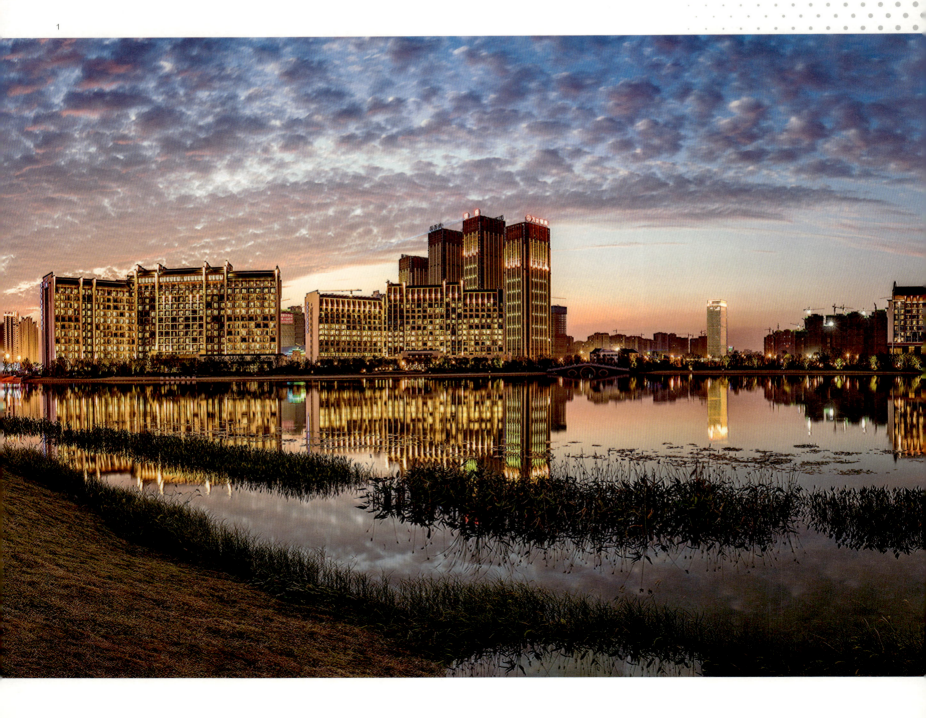

酒店群概况

合肥万达酒店群围绕人工湖布置酒店及酒吧街。全园的建筑形态与景观主题规划分为四个板块，分别为六星级酒店、五星级及四星级酒店群、三星级酒店群和酒吧街。酒店群地处巢湖之滨，拥有优美的自然景色和丰厚的徽州文化底蕴。总体设计以绿色生态为先、景观园林为主，将中式园林景观与徽派建筑空间有机结合，通过对建筑形态与景观朝向的匠心营造，创造出绿茵与水景融为一体的生态环境。建筑围绕湖景布局，力求做到"窗窗有景、处处见湖"，同时兼顾沿湖与沿街的立面形态以及城市空间的塑造。

通过微地形的营造产生出层次丰富、各具特色的庭院及滨水空间——六星级酒店以庭院和湿地景观为特色，建筑为低层院落式，兼顾湖景与内部庭院的景观，强调建筑环境的私密性与尊贵感；五星级及四星级酒店坐北朝南，为单廊半围合式，环抱开阔的沿湖叠落式田园景观；三星级酒店为双面客房的紧凑布局，拥有人工沙滩景观；酒吧街由一组形态丰富的村落式建筑构成，沿湖一侧为室外休闲茶座。

PROJECT OVERVIEW

The hotels and bar street are arranged around the artificial lake in Hotel Complex of Hefei Wanda. The building form and landscape planning in the Complex are divided into four categories: 6-star hotel, 5- and 4-star hotel complex, 3-star hotel complex and bar street. The hotel complex is next to the Chaohu Lake, where there are beautiful landscape and rich Huizhou culture. The master plan focuses on green ecology and landscape. It properly combines traditional Chinese style landscape with Hui Style architectural space. By working carefully with the architectural forms, landscape and orientation, an ecological environment is created where green land and water landscape are properly integrated. The buildings are planned around the lake landscape. "Beautiful landscape can be enjoyed at each and every window, and the lake can be seen everywhere" is the ultimate goal meanwhile the facade and urban space around the lake and along the street are considered.

By forming micro-topograph, courtyards and waterfront spaces with rich gradations and features are created - the 6-star hotel features courtyard and wet land landscape. The buildings are low-rise courtyard buildings. The lake landscape and courtyard landscape are taken care of. The building environment emphasizes privacy and noble sense. The 5-star and 4-star hotels are facing to the south with their back to the north, they are half enclosed by a corridor, offering open farmland landscape around the lake at different terraced levels. The 3-star hotels are in compact layout with rooms on both sides, and artificial sand beach can be enjoyed there. The bar street is composed of a group of village-like buildings with rich forms. On the side facing the lake, there are outdoor tables and seats for tea and drinks.

1 合肥万达城酒店群全景
2 合肥万达城酒店群总平面图

WANDA VISTA HEFEI
合肥万达文华酒店

OPENED ON: 24th, SEPTEMBER, 2016
LOCATION: BINHU DISTRICT, HEFEI, ANHUI PROVINCE
LAND AREA: 6.31 HECTARES
FLOOR AREA: 49,000 SQUARE METER
LANDSCAPE AREA: 48,900 SQUARE METER

开业时间：2016 / 09 / 24
开业地点：安徽省 / 合肥市滨湖区
占地面积：6.31 公顷
建筑面积：4.9 万平方米
景观面积：4.89 万平方米

5

3 合肥万达文华酒店立面图
4 合肥万达文华酒店入口牌楼
5 合肥万达文华酒店总平面图
6 合肥万达文华酒店入口

酒店概况

合肥万达文华酒店为六星豪华酒店，占地6.31公顷，建筑面积4.9万平方米，景观面积4.89万平方米，拥有客房或套房205间，为典型徽派风格。

PROJECT OVERVIEW

Wanda Vista Hefei is a 6-star luxury hotel, covering 6.31 hectares of land. The floor area is 49,000 square meters, and landscape 48,900 square meters. It has 205 rooms/suites and is designed with typical Hui Style.

6

设计理念

合肥万达文华酒店以庭院和湿地景观为特色,建筑设计为低层院落式布局,兼顾湖景与内部庭院的景观,强调建筑环境的私密性与尊贵感。设计将大堂及公共区、会议区和客房区分别组织成不同特色的庭院空间。设计运用粉墙、黛瓦、马头墙等"徽派建筑"的基本元素,取意传统徽州大院的厅、堂、廊、院,构成丰富多彩的空间层次,以黑、白、灰为主基调,在建筑的暗部点缀原木色,体现"青砖小瓦马头墙,回廊挂落花格窗"的徽派建筑特征。

DESIGN CONCEPT

Wanda Vista Hefei features courtyard and wet land landscape. The buildings are low-rise courtyard buildings. The lake landscape and courtyard landscape are both taken into account during the design process. The building environment emphasizes privacy and noble sense. The lobby, public areas, meeting areas and guest rooms are organized into featured courtyard spaces. In the design, whitewashed walls, black tiles, horsehead walls, and other fundamental elements used in Hui Style buildings are adopted. The halls, rooms, corridors and courtyards derives from the traditional Huizhou courtyard building, forming rich and colorful spatial gradations. The buildings are mostly in black, white and grey. In some dark shades of the buildings, the original colors of wood are used here and there to show the features of Hui Style buildings "black bricks, small tiles and horsehead walls, and along any winding corridor gridded windows can be found".

8

7 合肥万达文华酒店外立面
8 合肥万达文华酒店外立面
9 合肥万达文华酒店连廊

7

生态景观

合肥万达文华酒店的景观设计在尊重当地生态环境和传承当地文脉基础上，根据合肥当地独特的巢湖湿地自然条件及徽州的人文资源，将当地具有代表性的徽商文化、传说故事、建筑特色、材料工艺等纳入景观设计可利用资源的范畴；通过深刻的文化挖掘和艺术的空间规划，打造以"情归故里，梦回徽州"为主题的度假酒店景观。合肥万达文华酒店景观以"迎祥，圆梦，雅居"为主题线索。"迎祥"——亲友迎接徽州游子归家；"圆梦"——游子与家人团聚；"雅居"——营造具有素雅宁静之美，令人心驰神往之居所。景观以"三大主题"为线索，延伸出"万达文华十景"——荷风洗尘、歙砚归乡、雨雾茶田、四水归堂、金盏台、飞虹瀑、徽池雅调、画桥圆梦、翠竹幽境、清泉石上。结合宽阔的人工湖，设计湿地面积5000平方米，结合百米木栈道，种植水生植物3000平方米，共20个品种，实现水域自我净化的功能，并实现了酒店群良好的微环境。

ECOLOGICAL LANDSCAPE

While local ecologic environment is respected and local culture is inherited, the landscape design of Wanda Vista Hefei adopts the unique local natural conditions of wet land at the Chaohu Lake and the cultural resources of Huizhou, utilizing Hui commercial culture, legends, building features, materials, techniques, etc. for landscape design. By exploring the culture in depth and planning artistic space, the landscape of a holiday hotel is built under the subject of "returning to the hometown with love and going back to Huizhou in a dream". The landscape at Wanda Vista Hefei is designed on the theme of "Greeting auspice, fulfilling dreams and living gracefully". "Greet auspice" - The relatives and friends meet the travelers coming home. "Fulfilling dreams" - Travelers are reunited with their families. "Living gracefully" - Build plain, elegant and quiet homes that are loved by all. The landscapes are developed on "three subjects", and represents "Ten Beautiful Scenes of Wanda Vista" - Summer Wind Blows Dusts Away, She Inkstone Return Home, Rain and Fog at Tea Field, Rainwater from All Directions Collects at Courtyard, Jinzhantai, Rainbow Waterfall, Chizhou Opera of Anhui, Beautiful Bridge Fulfilling Dreams, Green Bamboos in Quietness and Clean Brook Running on Pebbles. Next to the vast artificial lake, the designed area of wet land is 5,000 square meters. Along the 100m wood boardwalk, 3,000 square meters of aquatic plants of more than twenty types are growing to purify the water and achieve a favorable micro-environment for the hotel complex.

10 合肥万达文华酒店湖景
11 合肥万达文华酒店水面栈道
12 合肥万达文华酒店景观
13 合肥万达文华酒店景观桥

WANDA REALM HEFEI
合肥万达嘉华酒店

OPENED ON : 24th SEPTEMBER, 2016
LOCATION : BINHU DISTRICT, HEFEI, ANHUI PROVINCE
LAND AREA : 4.2 HECTARES
FLOOR AREA : 38,000 SQUARE METERS
LANDSCAPE AREA : 28,900 SQUARE METERS

开业时间： 2016 / 09 / 24
开业地点： 安徽省 / 合肥市滨湖区
占地面积： 4.2 公顷
建筑面积： 3.8 万平方米
景观面积： 2.89 万平方米

酒店概况

合肥万达嘉华酒店为5星酒店，占地4.2公顷，建筑面积3.8万平方米，景观面积2.89万平方米，拥有客房或套房407间。

PROJECT OVERVIEW

Wanda Realm Hefei is a 5-star hotel, covering 4.2 hectares of land. The floor area is 38,000 square meters, and landscape 28,900 square meters. It has 407 rooms/suites.

14

15

14 合肥万达嘉华酒店总平面图
15 合肥万达嘉华酒店外立面
16 合肥万达嘉华酒店外立面
17 合肥万达嘉华酒店立面图

18 合肥万达嘉华酒店入口
19 合肥万达嘉华酒店外立面
20 合肥万达嘉华酒店门廊

设计理念

合肥万达嘉华酒店坐北朝南，以单廊围合环抱开阔的巢湖北岸，以层层递进式景观适应丰富的地形变化。酒店秉承对徽派建筑的忠实传承，利用传统徽派风格的马头墙、小青瓦、砖雕等建筑元素，营造出"白墙黛瓦花格窗"的徽派建筑形象；同时，高大而出挑的檐口、古朴典雅的色彩以及步移景异的室内外空间设计，使建筑既古韵浓郁又简洁精炼，完美诠释了徽州建筑的精华所在。

DESIGN CONCEPT

Wanda Realm Hefei is facing to the south with it's back to the north and is half enclosed by a corridor. The progressively transforming landscape is well-adopted to negotiate with the rich topography. The hotel faithfully follows the tradition of Huizhou architecture, and uses traditional Huizhou-style building elements like horsehead walls, small black tiles, brick carvings, etc. to create the Hui Style building with "white walls, black tiles and lattice windows". Large-sized cantilevers, plain and elegant colors and changing outdoor spacial design add ancient touches to the buildings while keeping them simple and refined, which is the perfect interpretation of the essence of Huizhou buildings.

生态景观

景观设计倡导绿色生态，体现花园景观与建筑空间的完美结合，创造出"绿茵共水景一色"的优美景致；将安徽的"山水印象"抽象提炼，以多样化的景观形式映射到场地中；通过叠石理水、植花理木，将徽州山水进行"复刻"，辅以路线回环，细节之中体现徽派风景之精粹。设置胸径30厘米以上大树200余棵，灌木1500余株，三季有花，四季常绿。

ECOLOGICAL LANDSCAPE

The landscape design advocates green and ecology, and perfectly integrates gardening landscape with architectural space to create the beautiful scene where "plants and water are in harmony". It abstracts the impression of mountains and rivers in Anhui, and shows them in the field in different landscapes. By using rockwork, water, flowers and trees, the landscape of Huizhou is copied. With the winding pathways, the detailed landscape shows the essence of Hui Style landscape. More than 200 large trees with the diameter over 30cm and 1500 shrubs are planted in the area. Flowers bloom in three seasons and the field is green throughout the year.

21 合肥万达嘉华酒店入口绿化
22 合肥万达嘉华酒店景观俯瞰
23 合肥万达嘉华酒店水景

PULLMAN HEFEI WANDA
合肥万达铂尔曼酒店

OPENED ON: 24th, SEPTEMBER, 2016
LOCATION: BINHU DISTRICT, HEFEI, ANHUI PROVINCE
LAND AREA: 3.18 HECTARES
FLOOR AREA: 38,000 SQUARE METERS
LANDSCAPE AREA: 21,000 SQUARE METERS

开业时间： 2016 / 09 / 24
开业地点： 安徽省 / 合肥市滨湖区
占地面积： 3.18 公顷
建筑面积： 3.8 万平方米
景观面积： 2.10 万平方米

酒店概况

合肥万达铂尔曼酒店占地3.18公顷，建筑面积3.8万平方米，景观面积2.10万平方米，拥有客房或套房413间。

PROJECT OVERVIEW

Pullman Hefei Wanda covers 3.18 hectares of land. The floor area is 38,000 square meters, and landscape area 21,000 square meters. It has 413 rooms/suites.

24 合肥万达铂尔曼酒店外立面
25 合肥万达铂尔曼酒店外立面
26 合肥万达铂尔曼酒店总平面图

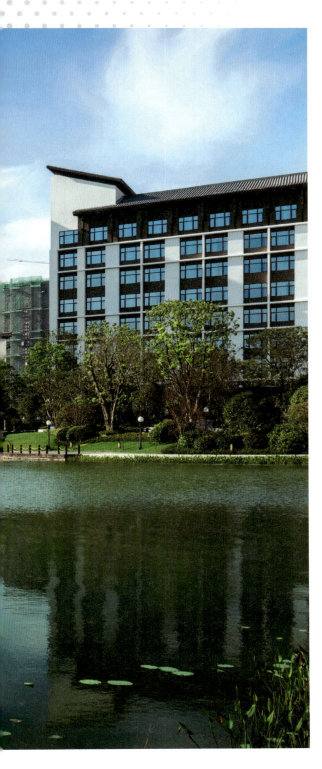

27

设计理念

合肥万达铂尔曼酒店在设计中简化提炼马头墙、木构架花窗、弯曲坡屋面等当地建筑元素，运用更具韵味的人字坡屋面和白墙来体现徽派风格。洁白的粉墙、黝黑的屋瓦、飞挑的檐角、鳞次栉比的兽脊斗拱以及高低错落、层层昂起的马头墙曲线，使得建筑外观别具美感。

DESIGN CONCEPT

The design of Pullman Hefei Wanda simplifies and abstracts horsehead walls, wood patterned windows, curved slope roofs and other local building elements, and represents Hui Style with charming gable roofs and white walls. Whitewashed walls, black tiles, overhanging eaves, rows upon rows of brackets, contour lines of horsehead wall of different heights make the buildings extraordinarily beautiful.

28

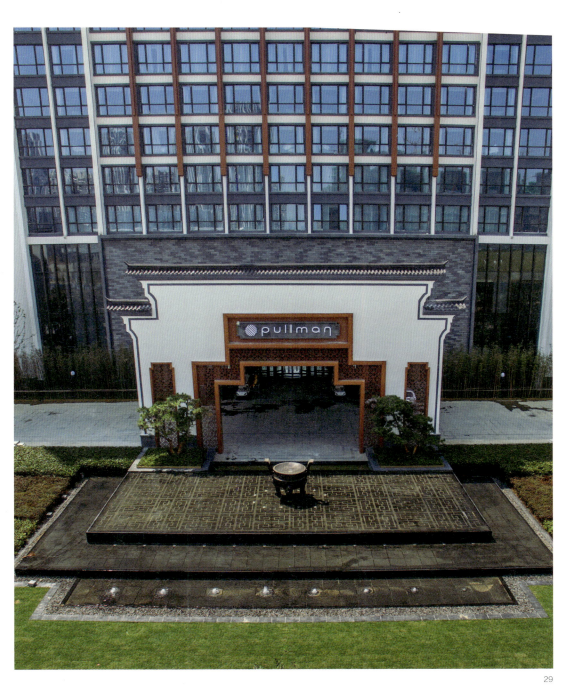

27 合肥万达铂尔曼酒店外立面
28 合肥万达铂尔曼酒店门廊
29 合肥万达铂尔曼酒店入口景观
30 合肥万达铂尔曼酒店水景

生态景观

采用当地原生树种，苗源地不超过200公里，减少"非苗木"移栽带来成活率低的问题。根据土质要求进行改良，大面积种植当地的茶树品种，配合雾喷设计，使得夏季园中小气候得到改良。

ECOLOGICAL LANDSCAPE

Original trees are planted as much as possible. The trees are purchased within 200 km from the hotel in order to maximize the survival rate. Improvements are made according to the quality of the soil. Local tea trees are planted in large areas. Working with the mist sprays, the temperature in hot summer will be lowered.

NOVOTEL HEFEI WANDA
合肥万达诺富特酒店

OPENED ON: 24th, SEPTEMBER, 2016
LOCATION: BINHU DISTRICT, HEFEI, ANHUI PROVINCE
LAND AREA: 2.79 HECTARES
FLOOR AREA: 38,000 SQUARE METERS
GROSS LANDSCAPE AREA: 23,900 SQUARE METERS

开业时间: 2016 / 09 / 24
开业地点: 安徽省 / 合肥市滨湖区
占地面积: 2.79 公顷
建筑面积: 3.80 万平方米
景观面积: 2.39 万平方米

酒店概况

合肥万达诺富特酒店位于合肥市滨湖区，用地面积2.79公顷，建筑面积3.80万平方米，景观面积2.39万平方米，拥有客房或套房589间。

PROJECT OVERVIEW

Novotel Hefei Wanda is located at Binhu District, Hefei. Its land area is 2.79 hectares, floor area 38,000 square meters, and landscape area 23,900 square meters, it has 589 rooms/suites.

31 合肥万达诺富特酒店总平面图
32 合肥万达诺富特酒店外立面
33 合肥万达诺富特酒店入口

设计理念

合肥万达诺富特酒店造型简洁朴实，采用古典的"三段式"构图，灵活运用传统马头墙分割建筑体量，将抽象简化的墙体穿插错落，弱化了大体量尺度而强化了地域特征，形成起伏的建筑轮廓线。酒店主入口落客区雨篷结合细致的木作与简洁的白墙——高大的马头墙，简化的黑色压顶石，灰砖勾边，石材勒脚——这些元素既传达出徽派建筑韵味又吐露现代气息。

DESIGN CONCEPT

The style of Novotel Hefei Wanda is concise and simple. It adopts the classic "tripartite composition" composition. The traditional horsehead wall is flexibly used to divide the buildings. The abstract, simple walls weaken the building scale and strengthen the regional features, forming an undulating building outline. The canopy over the drop-off area at the entrance of the hotel combines with delicate woodwork and simple white walls - large horsehead walls, simplified black cap stones, dark tile outline, stone footings - these elements express both the taste of Hui Style buildings and modern flavor.

MERCURE HEFEI WANDA
合肥万达美居酒店

OPENED ON: 24th, SEPTEMBER, 2016
LOCATION: BINHU DISTRICT, HEFEI, ANHUI PROVINCE
LAND AREA: 2.86 HECTARES
FLOOR AREA: 38,000 SQUARE METERS
LANDSCAPE AREA: 24,700 SQUARE METERS

开业时间: 2016 / 09 / 24
开业地点: 安徽省 / 合肥市滨湖区
占地面积: 2.86 公顷
建筑面积: 3.80 万平方米
景观面积: 2.47 万平方米

酒店概况

合肥万达美居酒店占地面积2.86公顷，建筑面3.80万平方米，景观面积2.47万平方米，拥有客房或套房577间。

PROJECT OVERVIEW

Mercure Hefei Wanda covers 2.86 hectares of land. The floor area is 38,000 square meters, and landscape area 24,700 square meters. It has 577 rooms/suites.

36

34 合肥万达美居酒店总入口
35 合肥万达美居酒店总平面图
36 合肥万达美居酒店大堂后场
37 合肥万达美居酒店外立面

37

设计理念

合肥万达美居酒店平面为L形，朝向湖面展开，沿湖一侧布置约70%的观景客房。首层的大堂、全日餐厅临湖设置，透过巨大的落地玻璃窗，优美景色一览无余。酒店采用传统的徽派建筑语言，用现代的材料及构成方式进行全新的阐释，营造出意蕴盎然的新中式徽派建筑之美。

DESIGN CONCEPT

With an L-shaped form, Mercure Hefei Wanda extends toward the lake surface, with about 70% of the rooms offering lakeside views. Both the lobby and the all-day dining at GF are facing the lake, so that guests could have a magnificent view of the beautiful scenes through the huge floor-to-ceiling glass windows. With traditional Hui style architectural language, the hotel adopts modern materials and composition manners for a brand-new interpretation, so as to create the architectural beauty in the new Chinese Huizhou style with profound implications.

04
BAR STREET OF HOTEL COMPLEX IN WANDA CITY HEFEI
合肥万达城酒店群酒吧街

OPENED ON: 24th, SEPTEMBER, 2016
LOCATION: HEFEI WANDA CITY, HEFEI
LAND AREA: 3.16 HECTARES
FLOOR AREA: 15,000 SQUARE METERS
LANDSCAPE AREA: 24,400 SQUARE METERS

开业时间： 2016 / 09 / 24
开业地点： 合肥市 / 合肥万达城
占地面积： 3.16公顷
建筑面积： 1.5万平方米
景观面积： 2.44万平方米

规划概述

合肥万达城滨河酒吧街是华东唯一的湖岸酒吧街，是"徽"元素与新潮流的文化碰撞，拥有16个品牌，包括酒吧、音乐餐吧、主题餐厅、咖啡、茶楼及KTV等丰富的业态。酒吧街呈前街后湖的"L"形布局。酒吧街沿街立面力求展现传统徽州街市的风貌，在路口转角处设置主入口广场，由状元坊、景观亭等传统徽派建筑要素形成的主轴线一直延伸到人工湖。在沿湖一侧结合景观设计布置了层次丰富的露台空间，把酒临风，体现现代都市繁华与传统文化优雅的结合。

PLANNING OVERVIEW

As the only lakeside bar street in East China, the bar street of Wanda City Hefei is a cultural collision between "Anhui" elements and the new trends; altogether 16 brands reside here, including bars, music cafes, theme restaurants, cafe, tea house, KTV and other various business types. The entire street is in an L-shaped layout with street in the front and lake at the back. The facade along the street focuses on showing the style and features of the traditional Huizhou street markets. At the road crossing a main entrance square is arranged; the main axle comprising of the Zhuangyuan workshop, landscape pavilion and other traditional Huizhou architectural elements extends till it reaches the artificial lake. Along the lakeside multiple layers of terrace spaces are designed in conjunction with the landscapes; against the wind with wine in hand, people can experience an elegant mix of the modern urban prosperity and traditional cultures.

3

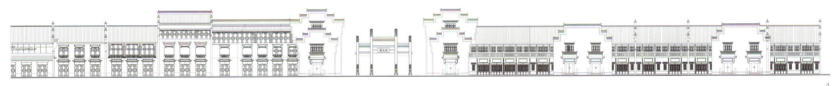

4

1 合肥万达城酒吧街鸟瞰图
2 合肥万达城酒吧街总平面图
3 合肥万达城酒吧街牌楼
4 合肥万达城酒吧街立面图
5 合肥万达城酒吧街建筑外立面

5

设计理念

酒吧街由"三段"不同风格的徽派建筑组成，以两层为主、局部三层。入口借经典徽州"牌坊"成为远处的状元楼之"框景"，再现古徽州"状元及第"的场景。酒吧街"后花园"以彰显徽州儒商文化的状元楼作为核心景点。酒吧街内街取题合肥具代表性的"庐阳八景"，以亲水台阶、绵绵荷叶作为辅助元素，烘托浓意水乡风貌。

DESIGN CONCEPT

The Bar Street is comprised of "three sections" of Hui-style architectures in different forms, mainly in two storeys and some in three. The entrance, with the classic Huizhou "memorial archway", becomes the "enframed scenery" of the distant Zhuangyuan Mansion and reproduces the scene of "Zhuangyuan passing the imperial examination" in ancient Huizhou. Zhuangyuan Mansion, which highlights the confucian entrepreneurship culture, is the core attraction of the "back garden" on the Bar Street. The representative "eight sceneries of Luyang" are used as the theme of the inner street of the Bar Street with water-adjacent steps and continuous lotus leaves as the supportive elements to set off the style and features of the water town.

WANDA PLAZA
万达广场

01

BEIJING FENGTAI HIGH-TECH PARK WANDA PLAZA
北京丰台科技园万达广场

OPENED ON: 22ⁿᵈ DECEMBER, 2016
LOCATION: FENGTAI DISTRICT, BEIJING
LAND AREA: 3.54 HECTARES
FLOOR AREA: 244,400 SQUARE METERS

开业时间: 2016 / 12 / 22
开业地点: 北京市 / 丰台区
占地面积: 3.54 公顷
建筑面积: 24.44 万平方米

1 北京丰台科技园万达广场鸟瞰图
2 北京丰台科技园万达广场总平面图
3 北京丰台科技园万达广场外立面

广场概述

北京丰台科技园万达广场位于北京市丰台科技园东区三期园区内，邻南五环路以及地铁9号线丰台科技园站，处于丰台科技园的重要节点位置。广场占地面积3.54公顷，总建筑面积24.44万平方米，包括大商业综合体、室外步行街、乙级写字楼及配套设施；其中大商业建筑面积19.18万平方米，室外步行街1.65万平方米，乙级写字楼3.29万平方米，城市配套设施0.32万平方米。

招商运营

北京丰台科技园万达广场商业定位"家庭品质生活+职场商务消费"为主线，"主攻"中端消费人群、年轻时尚人群和上班白领；引进国内外知名品牌260家，综合规划了服装、精品、餐饮、电子数码、娱乐、科技、体验和文化艺术等多种业态，为顾客带来丰富的购物乐趣和体验。

广场基于实体商业基础，植入互联网基因，全部品牌实现线下体验+线上消费，为品牌商家提供线上线下商业平台。开业以来推出了一系列丰富的主题福利活动：如"迎新盛典跨年狂欢夜"、"满月庆豪礼大放送"、"浪漫情人节"、"欢乐闹元宵"、"首届室内马拉松"、"十里桃花、心语心愿"、"CBA球星空降广场"等。在"快消费"时代依旧保留最真挚的本心，在繁闹的都市独守心中城堡。

PROJECT OVERVIEW

Beijing Fengtai High-tech Park Wanda Plaza is located in the Phase III compound of the East Zone of Beijing Fengtai Science Park, adjacent to the South Fifth Ring Road and Line 9 of Fengtai Science Park Station, Beijing Subway, which is an important node location of Fengtai Science Park. The Plaza covers an area of 3.54 hectares, with a gross floor area of 244,400 square meters, including large commercial complex, outdoor pedestrian streets, level B office building and supplementary facilities; the large commercial building covers an area of 191,800 square meters, the outdoor pedestrian streets 16,500 square meters, the level B office building 32,900 square meters, and city supplementary facilities 3,200 square meters.

INVESTMENT TENDER&OPERATION

The commercial orientation of Beijing Fengtai High-tech Park Wanda Plaza takes "quality family life + business & commercial consumption" as the mainline, "targeting mainly" at middle end consumers, fashionable young people and white-collar workers; it includes 260 domestic and foreign stationed brands, covering a wide range of business types (i.e. clothing, boutique, catering, digital products, entertainment, technology, experience and culture & art) to provide the customers with colorful shopping experience and fun.

The Plaza is based on traditional commercial entity and implanted with Internet genes to realize off-line experience + on-line consumption for all brands and provide online and off-line commercial platform for shops. Since its opening ceremony, a series of theme promotion activities are held, for example, "New Year's Rocking Eve Sale", "First Month Birthday Sale", "Romantic Valentine's Day Sale", "Happy Lantern Festival Sale", "First Indoor Marathon", "Ten Miles of Peach Blossoms, Heartfelt Wishes", "CBA Players in the Plaza", etc. We intend to remain the most sincere conscien-ce in an age of "fast consumption" and guard the castle of innocent heart in a busy metropolis.

4

5

广场外装

北京丰台科技园万达广场外立面以"电影胶片"为原型,通过连续的层间错动形成丰富的"界面表情",给人以一圈圈胶片缠绕而成的生动形象——主要材料是淡香槟色穿孔铝板和彩釉玻璃;两种材料通过交错布置,在立面上形成独具韵律感的视觉效果,以此隐喻以"电影文化"为脉络的设计概念——既以此建筑形象提升周边区域的"文化品质",也体现出"影视主题"在万达"类产业"中的主导地位。"本土设计"理念的运用使得建筑既与周边环境相融,又使得丰台科技园万达广场外立面造型卓尔不群。

FACADE DESIGN

The facade of Beijing Fengtai High-tech Park Wanda Plaza takes "filmstrip" as the prototype, forming rich "interface expression" via continuous strain-slip in layers, and impresses people with a vivid image of entwined filmstrips - main materials being light champagne perforated aluminum panel and enam-eled glass; through staggered arrangement of the two materials, a visual effect with unique rhythmi-cal image is formed in the facade, echoing the design concept of "film culture" - the image of the building not only increases the "cultural quality" of the surrounding areas, but also reflects the pre-dominance of "movie theme" in the "industries" of Wanda. The application of "native design" con-cept both allows the building to enter the surroundings and makes the facade of Fengtai High-tech Park Wanda Plaza rise above the common herd.

4 北京丰台科技园万达广场鸟瞰图
5 北京丰台科技园万达广场外立面
6 北京丰台科技园万达广场外立面
7 北京丰台科技园万达广场立面图

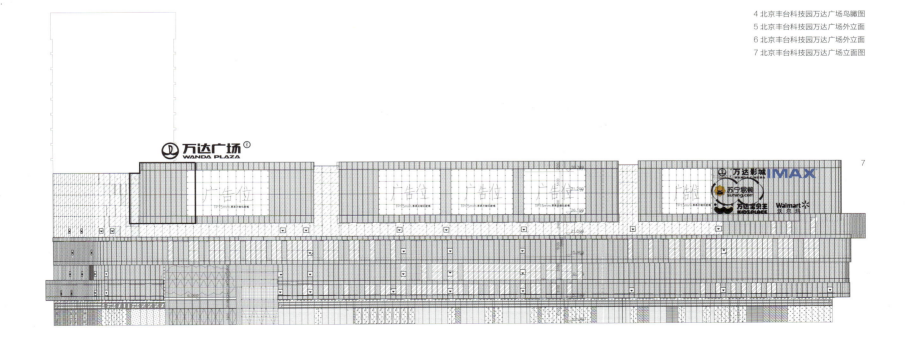

主力店　服装　精品　体验　儿童　餐饮

9 北京丰台科技园万达广场室内步行街
10 北京丰台科技园万达广场商业落位图
11 北京丰台科技园万达广场中庭
12 北京丰台科技园万达广场主入口顶棚
13 北京丰台科技园万达广场室内步行街

广场内装

北京丰台科技园万达广场内装方案以"向上的生命力"作为形态元素，将"科技"与"绿色"理念相结合，使得购物中心成为人性、智能、节能环保的建筑。其装饰手法以"植物强有力地扎入泥土，体现不断向上伸展的力量"为设计理念，使得整个购物中心不仅富有力量与气魄，还体现了大自然的无限生机律动。

INTERIOR DESIGN

The interior design scheme of Beijing Fengtai High-tech Park Wanda Plaza takes "progressive life force" as its elements, combines the concepts of "technology" and "green", making the shopping center a hu-mane, intelligent, eco-friendly building. The decoration takes "plunging plants forcefully into the earth to reflect continuous, progress and extensive force" as the design concept, rendering the whole shopping mall not only with strength and boldness of vision but also with the infinite vitality and rhythm of the nature.

14

广场景观

北京丰台科技园万达广场景观设计灵感源自"LINE"（网络），用抽象线条演绎网络空间，在现实中展现虚拟的网络空间，造就具有科技感、时尚感的广场氛围。设计充分发挥场地本身（滨河景观）的地理优势，打造集亲水游玩及休憩于一体的商业景观街区，突出城市广场的生态性。景观雕塑结合北京地域文化，从传统京剧人物形象中抽象衍化，搭配现代著名油画图案，实现传统与现代的完美融合。

LANDSCAPE DESIGN

The landscape design inspiration of Beijing Fengtai High-tech Park Wanda Plaza derives from "LINE" (internet), expressing cyberspace with abstract lines, revealing virtual cyberspace in reality, and creating an atmosphere with sense of technology and fashion in the plaza. The design gives full play to the geological advantage of the site itself (the riverside landscape) and creates a commer-cial landscape block for riverside sightseeing and recreation, highlighting the ecology of urban plaza. The landscape sculptures combines the regional culture of Beijing, deriving from the im-ages of traditional figures in Peking Opera and matched with patterns of famous modern painting, so as to realize a perfect fusion of the traditional and the modern.

15

16

广场夜景

北京丰台科技园万达广场夜景照明充分展现并提升建筑"电影胶片"的设计理念，实现亮丽舞动的夜景效果——穿孔板内透出的立面光与方块灯构成"胶片形象"；吊顶边沿的光带塑造"胶片"的错层关系；入口运用装饰照明结合投光，突出建筑入口的醒目性。照明与建筑的"一体化"设计，使建筑从环境里脱颖而出，具有独特的艺术性。

NIGHTSCAPE DESIGN

The nightscape lighting of Beijing Fengtai High-tech Park Wanda Plaza fully reflects and elevated the design concept of "filmstrip", realizing a bright and sparkling nightscape effect - the facade light from per-forated plates and square lights constitute "filmstrip images"; the light bands along the edges of the ceiling create a staggered effect; decoration lighting and project lighting are used at the entrance to highlight the entrance of the building. The "integrated system" design of the lighting and the build-ing, with its unique artistry, help the building stand out from the surrounding buildings.

14 北京丰台科技园万达广场景观小品
15 北京丰台科技园万达广场景观小品
16 北京丰台科技园万达广场景观小品
17 北京丰台科技园万达广场夜景
18 北京丰台科技园万达广场夜景

02
BEIJING HUAIFANG WANDA PLAZA
北京槐房万达广场

OPENED ON : 22nd DECEMBER, 2016
LOCATION : FENGTAI DISTRICT, BEIJING
LAND AREA : 7.12 HECTARES
FLOOR AREA : 230,000 SQUARE METERS

开业时间：2016 / 12 / 22
开业地点：北京 / 丰台区
占地面积：7.12 公顷
建筑面积：23.0 万平方米

2

1 北京槐房万达广场外立面
2 北京槐房万达广场总平面图

1

PART C WANDA PLAZA 万达广场

广场概述

北京槐房万达广场位于北京市丰台区槐房新村,临近规划南环公路,距离地铁4号线新宫地铁站仅50米,是连接城区的重要节点;坐拥新机场线的交通大动脉格局,是未来城区的交通核心。广场共四层,总规划用地面积7.12公顷,总建筑面积23.0万平方米,包括购物中心、室外步行街、地下车库及设备用房等配套设施。其中大商业建筑面积18.87万平方米,室外步行街建筑面积3.15万平方米,地下车库建筑面积0.93万平方米,城市配套物业建筑面积500平方米。

招商运营

北京槐房万达广场立足于提升"京南"消费品位、完善城市区域功能,打造"体验感好、时尚感强、知名度高"的时尚地标、购物圣地。广场汇聚了316个国内外知名品牌,引进全球领先的万达影城、大型室内儿童娱乐主题乐园——万达宝贝王、全球领先的连锁电玩城——大玩家,以及永辉超市在内的实力连锁超市和"十大品牌"主力店,真正实现了购物、休闲、娱乐的"一站式"生活新体验,打造京南新地标!

PROJECT OVERVIEW

Beijing Huaifang Wanda Plaza s located in Huaifang New Residential Quarter, Fengtai District, Beijing, adjacent to the South Ring Road under planning and only 50m away from Line 4 of Xingong Station, Beijing Subway, which is an important node location that connects to the downtown area; it's on the route of the new airport line and would be a traffic core of the urban area in the future. The Plaza has four floors, with a site area of 7.12 hectares and a gross floor area of 230,000 square meters, including shopping mall, outdoor pedestrian street, underground parking garage, plant rooms and other supplementary facilities; the large commercial building covers an area of 188,700 square meters, the outdoor pedestrian streets 31,500 square meters, the underground parking garage 9,300 square meters and city supplementary facilities 500 square meters.

INVESTMENT TENDER & OPERATION

Beijing Huaifang Wanda Plaza a ms to elevate the consumption taste of "South Beijing", to perfect the regional function of the city and to create a fashionable landmark and shopping area that provides great "experience, fashion sense and popularity". The Plaza has 316 foreign and domestic brands, including the advanced Wanda Cinema, large indoor children's theme park - Wanda Kidsplace, world-leading chain game center - Big Player, competent chain supermarkets (i.e. Yonghui Superstores) and anchor stores of "Top Ten Brands", truly realizing the purpose of providing "one-stop" new life experience covering shopping, recreation and entertainment and constructing a new landmark in "South Beijing".

3

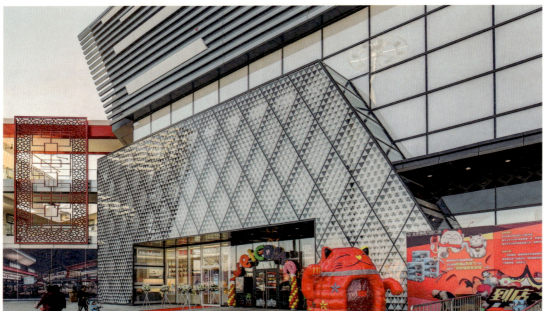
4

6

广场外装

北京槐房万达广场外立面以"飞翔"为设计理念,通过时尚灵动的线条和精致优美的细部勾勒出整体动感;不同尺度、颜色的金属铝材层层搭叠,构建流动的层次感;倾斜状变化的"飞翼"收边更是强化了展翅飞翔的设计初衷。设计于流线型线条中加入了不同材质的"盒子",使不同体块之间穿插交接,赋予立面更多的精彩变化——在形成商业氛围的同时,"飞翔"的形体既体现了北京国际化大都市的"精气神",更是对万达不断前行、不断拼搏的企业精神的诠释。

FACADE DESIGN

The facade of Beijing Huaifang Wanda Plaza takes "flying" as the design concept, and the overall dynamic of the building is outlined through the sleek lines and delicate & exquisite details; the multi-layered overlapping of different scales and colors construct a flowing sense of layer; the "flying wing" edge of tilting change enhances the original design concept - spread its wings to fly. The design adds "boxes" of different materials into the linear lines, realizing the transition and interweaving of different masses and rendering more wonderful changes to the facade - while forming the commercial atmosphere, the form of "flying" is not only an embodiment of the "soul and spirit" of Beijing as an international metropolis, but also an interpretation of the enterprise spirit of Wanda - keep moving forward and striving continuously.

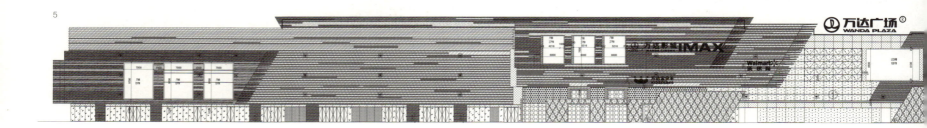
5

3 北京槐房万达广场外立面
4 北京槐房万达广场主入口
5 北京槐房万达广场立面图
6 北京槐房万达广场外立面
7 北京槐房万达广场外立面

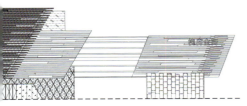

广场内装

室内设计灵感取材自"宇宙空间站"——入口顶棚使用白色铝板分割,隐喻宇宙空间站顶棚;中庭的造型取自宇宙飞船的形状;色调主要使用白色、黑色和灰色,也与宇宙空间站的主题相对应。

INTERIOR DESIGN

The interior design is inspired by the "space station" - the ceiling at the entrance adopts white aluminum plates, which is a metaphor for the space station's ceiling; the atrium takes the shape of a spaceship; the tones are mostly white, black and grey, corresponding to the theme of the space station.

8 北京槐房万达广场中庭
9 北京槐房万达广场室内步行街电梯
10 北京槐房万达广场室内步行街连桥
11 北京槐房万达广场商业落位图

PART C　WANDA PLAZA
万达广场

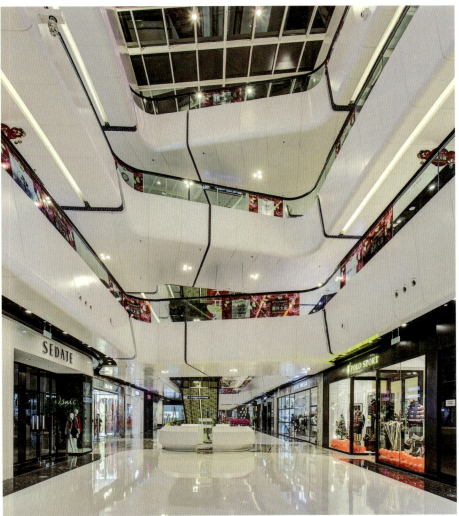

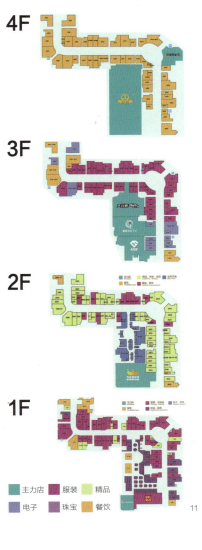

4F
3F
2F
1F

主力店　服装　精品
电子　珠宝　餐饮

广场景观

北京槐房万达广场以北京当地特色文化为主，以"枝言片语"为主题——国槐的"枝"寓意繁荣昌盛，塑造一个生机盎然的购物公园；"片语"从京剧及民俗生活语言中，挖掘文脉肌理，传承地域文化，演绎正统"京味视觉"。景观雕塑延续"北京文化"的设计主题，提取具有"老北京"特色的代表性元素，如国槐、冰糖葫芦、京剧等元素，结合现代装饰手法进行抽象提炼；在现代装饰艺术中融入传统文化内涵，深切表达"老北京"情怀。

LANDSCAPE DESIGN

The landscape design of Beijing Huaifang Wanda Plaza is dominated by the local characteristics of Beijing, with the theme of "branch language and isolated phrases" - the "branch" of the Chinese locust tree symbolizes prosperity, creating a vibrant shopping park; "isolated phrases" excavate the vein texture from Peking Opera and folkloric life language, inherit the regional culture and show the orthodox "Beijing-style Vision". The landscape sculptures also adhere to the design concept of "Beijing Culture". Representative elements that are rich in the characteristics of "Old Beijing" (i.e. Chinese scholar tree, sugar-coated haws on a stick, Peking Opera) are extracted and applied in combination with modern decoration techniques; the connotation of traditional culture is mixed into the art of modern decoration, expressing a profound romanticism of "Old Beijing".

12

13

14

15

16

12 北京槐房万达广场绿化
13 北京槐房万达广场景观
14 北京槐房万达广场绿化
15 北京槐房万达广场夜景
16 北京槐房万达广场夜景

广场夜景

夜景照明注重建筑夜间氛围的表达，强化建筑"体块关系"的对比，并通过不同材质与颜色的差别体现建筑的丰富感——在"铝方通"的"羽翼"位置内置导光板，表达建筑飞翔的寓意；立面"铝方通"凹槽安装线性洗墙灯，"见光不见灯"，轻盈灵动；入口"玻璃盒子"晶莹剔透，简单大气；黄金色铝板与银色铝板之间用大功率洗墙灯，强化不同体块的分界关系。

NIGHTSCAPE DESIGN

Nightscape lighting pays attention to the expression of building atmosphere at night, strengthens the building "massing" comparison and reveals the richness of the building through different materials and colors - built in light guides are installed in the "wing" position of the "aluminum rectangular tube" to express the implication of flying building; linear wall washer lights are installed in the grooves of the facade "aluminum rectangular tube" to achieve the vivid effect of "seeing light without light"; the "glass boxes" at the entrance are glittering, translucent, simple and elegant; high-power wall washer lights are installed between golden aluminum plates and silver aluminum plates to strengthen the division of different blocks.

03

CHENGDU SHUDU WANDA PLAZA
成都蜀都万达广场

OPENED ON : 29ᵗʰ APRIL, 2016
LOCATION : CHENGDU, SICHUAN PROVINCE
LAND AREA : 17.4 HECTARES
FLOOR AREA : 688, 900 SQUARE METERS

开业时间： 2016 / 04 / 29
开业地点： 四川省 / 成都市
占地面积： 17.4 公顷
建筑面积： 68.89 万平方米

1 成都蜀都万达广场外立面
2 成都蜀都万达广场总平面图
3 成都蜀都万达广场概念草图

1

广场概述

成都蜀都万达广场位于成都市郫都区，紧邻成灌快铁郫县站，是通往世界著名风景名胜区都江堰、青城山、黄龙和九寨沟的必经之路。广场占地17.4万平方米，总建筑面积68.89万平方米。广场由购物中心、室外步行街、SOHO公寓和住宅组成，其中购物中心建筑面积10.03万平方米，室外步行街8.16万平方米，SOHO公寓建筑面积17.62万平方米，住宅建筑面积32.92万平方米，其他配套类面积0.16万平方米。

招商运营

蜀都万达广场是郫县首位城市综合体，也是万达集团布局成都的第三座万达广场，作为万达集团第3.5代体验式商业综合体的升级版，代表着国内商业综合体的新水准。广场聚合近千品牌商家，以"吃喝玩乐一体化"体验式消费，颠覆成都城西商业格局；以17.7万平方米的体量汇聚13家主力店，如万达影院、大玩家、宝贝王、永辉超市、优衣库、苏宁云商、威斯特健身、火上芭蕾主题餐厅等。

在运营方面，广场举办"年中庆"活动，举办12场大型高空机车秀，配合"音乐狂欢节"打造"郫县不夜城"；活动期间召集全城"萌宝"进行"才艺大比拼"，得到广泛影响；活动现场特制50米巨型蛋糕，把节日的喜庆与顾客分享，成为全城参与的消费盛事。

PROJECT OVERVIEW

Chengdu Shudu Wanda Plaza is located in Pixian County, Chengdu, adjacent to Chengguan Express Railway, which is a spot on the only way to internationally known scenic spots -Dujiangyan, Mount Qingcheng, Yellow Dragon Scenic Area and Jiuzhaigou Valley. The Plaza covers an area of 174,000 square meters, with a gross floor area of 688,900 square meters. The Plaza consists of shopping mall, outdoor pedestrian streets, SOHO apartments and residential buildings; the shopping mall covers an area of 100,300 square meters, the outdoor pedestrian streets 81,600 square meters, the SOHO apartments 176,200 square meters, the residential building 329,200 square meters and other supplementary facilities 1,600 square meters.

INVESTMENT TENDER & OPERATION

Shudu Wanda Plaza is the first city complex in Pixian County and the third Wanda Plaza in Chengdu. As an updated version of Wanda Group's 3.5 generation experiential commercial complex, it represents the new level of domestic commercial complex. The Plaza has almost 1000 brands, subverting the business pattern of west Chengdu with experimental consumption of "eating, drinking, playing and entertaining all at one-stop"; with an area of 177,000 square meters, it gathers 13 anchor stores, including Wanda Cinema, Super Player, Wanda Kidsplace, Yonghui Superstores, Uniqlo, Suning Appliance, Sister Fitness and Water Ballet.

In terms of operation, the Plaza has held a "mid-year promotion" activity, 12 large overhead bike shows and a "music carnival" to make Pixian County into a "sleepless city"; during the event, the "cute babies" in the whole city are convened to "have a talent competition", which has received great attention from the public; a specially-made gigantic 50-meter cake is brought to the event site to share the happiness of the festival the customers, which becomes a great consumption event with the participation of the whole city.

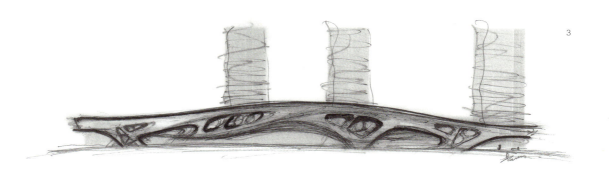

4

4 成都蜀都万达广场主入口
5 成都蜀都万达广场立面图
6 成都蜀都万达广场外立面
7 成都蜀都万达广场侧立面

5

广场外装

广场建筑整体色调以银白色为主,外轮廓采用流动、舒展的线条,以优美的连续起伏的韵律向东西两个方向伸展,如水波起伏、荡开,优雅又动感十足,体现着速度、活力和奔放的热情,具有强烈的时代精神。外立面以铝板幕墙为主,表皮分为三个层次,三层曲线在各自不同的维度和轨道上游弋穿梭,形成动感有机、虚实交替的图底关系。入口延续了建筑表皮的"层化",在转角面上塑造完整流畅的弧形曲线并由外层表皮逐层渐次收进,一方面以"展开"的方式强调了入口的位置与形态,一方面勾勒出入口处"洞"与"天"的意象。

FACADE DESIGN

The overall tone of the Plaza is dominated by silver white, and the outer contour adopts smooth and stretched lines, extending ups and downs to the wes: and the east in a graceful and continuous rhythm, like the waving and swinging of water with elegance and dynamic, embodying speed, dynamic and bold enthusiasm and a strong spirit of the age. The facade is dominated by aluminum curtain wall with the three layers. The three layers of curves moves through different dimensions and orbits, forming a dynamic, organic and void-solid alternative figure-ground relation. The entrance continues the "stratification" of the facade, shaping complete and smooth arc curves on the corner in a gradual way layer by layer from the outer layer. On the one hand, it emphasizes the position and form of the entrance by "stretching" ; on the other hand, it outlines the image of "the hole" and "the sky" at the entrance.

8

9

10

11

8 成都蜀都万达广场入口顶棚
9 成都蜀都万达广场中庭
10 成都蜀都万达广场室内步行街侧裙
11 成都蜀都万达广场商业落位图
12 成都蜀都万达广场中庭

广场内装

设计元素取材于历史文化符号——蜀绣。设计以"针织"作为理念，以"整齐针脚、虚实合度"作为空间形态风格，如侧裙GRG的横条纹造型、连桥扶梯的立体塑造、灯光的线面效果以及图案的运用，都充分融合了蜀绣的针织技法和蜀绣的图案特点，使平面效果和立体空间变得活泼生动富有节奏感，具有一气呵成、气韵连贯的艺术效果。

INTERIOR DESIGN

Chengdu Shudu Wanda Plaza draws design elements from the historical and cultural symbols - Shu embroidery. The design takes "knitting" as a concept, with a spatial format style of "neat stitching and void-solid cooperation". The side GRG horizontal stripes modeling, the three-dimensional modeling of the bridge escalator, the linear and plane effect of lighting and the application of patterns fully integrate the knitting technique and pattern characteristic of Shu embroidery, making the plane effect and the three-dimensional space vivid and rich in rhythm, with a coherent and artistically consistent effect.

13

14

广场景观

成都蜀都万达广场景观设计以"天府蜀韵"为设计灵感,运用动感而连续的曲线作为平面设计主题来打造地面铺装及相关小品和种植池,起始于各个入口的曲线在广场中央会合,既与建筑立面相融合,又象征着"历史长河"。在"历史长河"中提取"杜宇化鹃"、"三国蜀汉"、"川剧"等文化珍宝符号,并将其提炼成景观小品,再放置回现实场景,具有超凡脱俗的韵味。

LANDSCAPE DESIGN

The landscape design of Chengdu Shudu Wanda Plaza is inspired by "self-sufficient region with the features of Shu", using dynamic and continuous curves as graphic design theme to shape the floor paving and related sketches and planting pool. The curves starting from each entrance meet in the center of the Plaza, which not only integrate with the building facade but also embody the "long river of history". Valuable cultural symbols (i.e. the Kingdom of Shu-han and Sichuan Opera) are extracted from the "long river of history", made into landscape sketches and put back into reality scenes, which embody extraordinary and lasting appeal.

13 成都蜀都万达广场景观花坛
14 成都蜀都万达广场景观雕塑
15 成都蜀都万达广场绿化
16 成都蜀都万达广场夜景
17 成都蜀都万达广场室外步行街

15

室外步行街

蜀都万达金街打造基于对老成都传统民居的建筑风格和传统建筑材料的传承，借鉴宽窄巷、锦里等老成都风情商业街的造型元素和装饰构件，打造既有文化古韵又现代时尚的商业街。以牌坊、门楼等古典醒目的造型重点打造金街出入口、丁字交叉口等关键节点，引人进入、驻足。坡顶、挑檐穿插轻钢龙骨平顶雨篷，青砖青瓦与彩釉玻璃、金属栏杆扶手相得益彰，营造出"一店一色"。室外步行扶梯和玻璃观光梯，带动一层和二层外廊平台的交织互动，取得步移景异的效果。

OUTDOOR PEDESTRIAN STREET

The construction of the Gold Street of Shudu Wanda is based on the inheritance of the architectural style and traditional construction materials of traditional residential houses in old Chengdu. With reference to the modeling elements and decoration members of old Chengdu commercial streets (i.e. china line and Jinli), a commercial street with both cultural and ancient charm as well as modern appeal is constructed. Classic and striking modelings (i.e. memorial gates and arches over gateways) are adopted to construct key nodes, like the entrances and exits of the Gold Street and the three-legs intersection, attracting people to enter and stay. Light-steel keel flattop canopy, green tiles and bricks, enameled glass and metal railings interweave on the sloped roofs and cornices, echoing each other and creating an effect of "One Shop, One Style". The outdoor pedestrian elevator and glass sightseeing lift bring along the integration and interweaving of the veranda platform on the first and second floor, achieving the effect of varying sceneries with changing view-points.

04

JI'NAN HIGH-TECH WANDA PLAZA
济南高新万达广场

OPENED ON: 18th JUNE, 2016
LOCATION: JI'NAN, SHANDONG PROVINCE
LAND AREA: 4.87 HECTARES
FLOOR AREA: 171,800 SQUARE METERS

开业时间： 2016 / 06 / 18
开业地点： 山东省 / 济南市
占地面积： 4.87 公顷
建筑面积： 17.18 万平方米

1

广场概述

作为济南高新区最大的城市综合体，济南高新万达广场位于济南市高新区会展中心东侧，南邻工业南路，东侧为颖秀路，集购物中心、室外商业步行街及商铺、商务公寓、五星级酒店于一体。其中，购物中心属于高层商业建筑，近9万平方米，地上6层（顶层为影院夹层）、地下两层，由娱乐楼和室内商业步行街、地下商业及停车库组成，包括超市（位于地下一层）、影院、精品店及餐饮区、健身房、儿童天地等业态。

招商运营

济南高新万达广场是济南高新区第一个真正意义的生活体验式购物中心，是一座集购物、餐饮、文化、娱乐、休闲等多种业态于一体的城市综合体，为附近居民提供了很大的购物便利。这座济南高新万达广场，以"泉城印象"——山东首条室内特色文化主题餐饮街区而闻名。"泉城印象"集合了时尚主题餐厅、地方特色餐厅、国内知名餐饮、大型连锁餐饮、甜品小食等各类美食，引领城市消费风潮，全面提升济南人民的高品质生活体验。

PROJECT OVERVIEW

As the biggest city complex in the High-tech District of Ji'nan, the Ji'nan High-tech Wanda Plaza is located to the east of the Ji'nan High-tech District Convention and Exhibition Center, to the north of Industry North Road and to the west of Yingxiu Road. The plaza consists of a shopping mall, outdoor pedestrian streets, apartments, commercial apartments and a five-star hotel. The shopping mall is a high-rise commercial building, covering an area of about 90,000 square meters. It has 6 floors above-ground (the top floor is an interlayer for cinema) and 2 floors underground. It consists of an recreation building, indoor pedestrian streets, underground commerce and parking garage, including business types of supermarkets (located on B1), cinema, boutiques, catering area, fitness center, playgrounds for children, etc..

INVESTMENT TENDER & OPERATION

Ji'nan High-tech Wanda Plaza is the first true life experience-style shopping center in Ji'nan High-tech District. It is a city complex covering the business types of shopping, catering, culture, entertainment and recreation, providing convenient for the surrounding residents, who no longer need to "cross mountains and river" to go shopping. Ji'nan High-tech Wanda Plaza is famous for the "Impression of Spring City" - the first indoor featured culture-themed catering block in Shandong Province. "Impression of Spring City" gathers fashionable themed restaurants, local featured restaurants, domestically renowned restaurants, large chain restaurants, deserts and snacks. It is the trendsetter of consumption in Ji'nan and comprehensively elevates the high-quality life experience of citizens in Ji'nan.

PART C　WANDA PLAZA
万达广场

1 济南高新万达广场总平面图
2 济南高新万达广场外立面
3 济南高新万达广场立面图

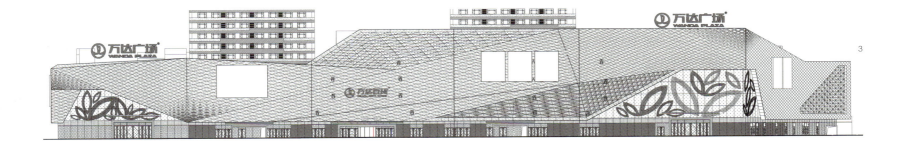

4 济南高新万达广场外立面
5 济南高新万达广场外立面
6 济南高新万达广场外立面

广场外装

济南"四面荷花三面柳，一城山色半城湖"，以泉水、荷花、柳叶久负盛名。设计从千佛山倒映在大明湖水面的具象中提取灵感，形成了"明湖天地，佛山倒影"的概念。广场形象提取自山体刚劲有力、连绵起伏的线条，形成气势恢宏、交错转折的形体轮廓。在理念上，在"大实"的"山体形态"中穿插"大虚"的"湖面印象"相呼应。在材料上，形体"大实面"采用亮银色铝板，表达时尚、现代、未来感；"大虚面"采用镜面玻璃和彩釉玻璃，表达大明湖波光粼粼的意向。于是，形成"虚实"相映，厚重与灵动并蓄的立面效果。

FACADE DESIGN

As an old saying goes, Ji'nan is a city with "four sides of lotus flowers and three sides of willows, a city of mountain view and a half city of lakes". Ji'nan has been famous for its spring water, lotus flowers and willow leaves for a long time. The design is inspired by the reflection of Mount Qianfo on the surface of Daming Lake, forming the concept of "the universe of Gaming Lake and the reflection of Mount Qianfo". The image of the Plaza is drawn from the powerful, vigorous, rolling and continuous lines of the mountain, forming a magnificent and staggered contour. In concept, the "solid" format of the mountain echoes with the "void" impression of the lake surface. As for materials, the "solid surface" of the form adopts bright silver aluminum plate to embody the sense of fashion, modern and future; the "void surface" adopts mirror glass and enameled glass to express the image of the glittering Gaming Lake. Therefore, the facade achieves an effect of "solid-void" contrast, with both sense of heaviness and lightness.

PART C WANDA PLAZA
万达广场

广场内装

万达广场的室内设计概念为"泉之韵",以简洁的线条和造型元素营造诗意盎然的商业氛围。设计摒弃了繁复的装饰语言,强调设计的本源味道和纯粹性——在采光顶以及地面铺装上,突出展示流动感和无序性,以曲线白描的手法诉说济南泉城之泉水元素;侧裙部分以金色光芒勾勒空间的轮廓。在这座"凝固的音乐"中唱响泉水之韵律,吐露天地之灵气。

4F

3F

2F

1F

■ 主力店 ■ 服装 ■ 精品 ■ 体验 ■ 餐饮

9

7 济南高新万达广场中庭
8 济南高新万达广场中庭侧廊顶棚
9 济南高新万达广场商业落位图
10 济南高新万达广场室内步行街连桥

10

INTERIOR DESIGN

The interior design concept of Wanda Plaza is "the rhyme of the spring", creating a poetic commercial atmosphere with simple lines and modeling elements. The design abandons complicated decorations and stresses the original purpose and purity of design — for the skylight ceiling and ground pavement, the design stresses the expression of mobility and disorder, elaborating the element of spring in spring city Ji'nan with the technique of line drawing in traditional ink and brush style; golden light is used to outline the spatial contour of the side skirt. This Plaza of "solidified music" rings with the rhythm of spring and emits the power of the universe.

11

广场景观

在景观设计上，将具有泉城文化代表性的"白鹭"融入其中，仿佛白鹭盘旋于山顶，俯瞰济南的古往今来，体会济南的风土人情。设计将白鹭羽翼的优美舞姿体现在景观中，演变成广场空间的要素，展现济南美景。广场景观融合地域文化元素，打造商业广场文化特质新气象，传承济南文化脉络，提升和带动区域环境品质，成为区域发展的催化剂。

LANDSCAPE DESIGN

The "Egret", a representative element of the culture of the Spring City, is integrated into the landscape design of the Plaza, as if an Egret is hovering on the top of the mountain, overlooking the city of Ji'nan through all ages and experiencing the local conditions and customs of Ji'nan. The design shows the elegant dancing posture of Egret wing in the landscape, which becomes an important element of the Plaza to reveal the beautiful scenery of Ji'nan. Local cultural elements are integrated into the plaza landscape to create the cultural new look of the commercial plaza, to inherit the cultural context of Ji'nan and to relevant and motivate the regional environment quality, becoming the catalyst for regional development.

12

13

广场夜景

济南是名副其实的山水城市。广场夜景灯光以山水为灵感，以幕墙立面为载体，突出建筑体量感。塔楼灯光以线条与面光为主，凸显夜间建筑的挺拔之姿，象征巍峨的山体；商业中心灯光以"水"为元素；主要强调整个建筑的几何形体，勾勒不规则菱形线条，呈现出不同氛围的灯光动画形式；立面三角形彩釉玻璃忽明忽暗、若隐若现，宛若大明湖波光粼粼的水面；线条之流畅与点之灵动遥相呼应，使建筑在夜间熠熠生辉。

NIGHTSCAPE DESIGN

Ji'nan is a veritable city with great landscape views. The nightscape lighting design of the Plaza is inspired by the mountains and rivers. The facade of the curtain wall is taken as the carrier to highlight the dimension sense of the building. The lighting of tower building mainly adopts linear light and ceiling spotlight to highlight the towering figure of the building at night, embodying the lofty mountain; the light of the commercial center takes "water" as an element, stressing the geometric form of the whole building and outlining the anomaly rhombic lines to show the forms of light settings under different atmospheres: the triangle glazing glass on the facade flickers and appears indistinctly like the glittering surface of the Gaming Lake; the liquidity of the lines and the mobility of the spots each other at a distance, making the building sparkling at night.

11 济南高新万达广场景观
12 济南高新万达广场景观雕塑
13 济南高新万达广场景观小品
14 济南高新万达广场夜景
15 济南高新万达广场室外步行街

室外步行街

济南拥有丰富文化底蕴，设计沿袭"老济南"特色建筑风格，例如"德华银行旧址、老济南站、宏济堂、将军庙、天主教堂、老商埠、邮政局旧址"等，从中选取典型的建筑语言和代表性建筑构件，与简欧式的设计风格相结合，诠释全新的商业建筑文化。

OUTDOOR PEDESTRIAN STREET

Ji'nan has a rich cultural background, so the design inherits the featured architectural style of "Old Ji'nan", for example the "former site of Dehua Bank, the former Jinan Station, Hong Ji Tang, the General's Temple, the Cathedral, the old commercial port and the former site of the Post Office". Architectural language and representative construction members are drawn from the above buildings and combined with simple-European design style to show a brand new culture of commercial building.

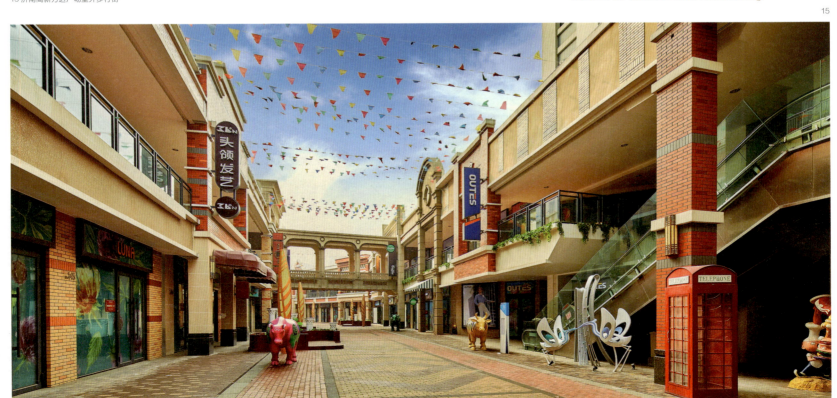

05
YANTAI DEVELOPMENT ZONE WANDA PLAZA
烟台开发区万达广场

OPENED ON: 23rd DECEMBER, 2016
LOCATION: YANTAI, SHANDONG PROVINCE
LAND AREA: 4.87 HECTARES
FLOOR AREA: 130,000 SQUARE METERS

开业时间：2016 / 12 / 23
开业地点：山东省 / 烟台市
占地面积：4.87公顷
建筑面积：13.0万平方米

广场概述

烟台开发区万达广场位于烟台经济技术开发区，衡山路以西、沭河路以南、淮河路以北，占地4.87公顷，总建筑面积13.0万平方米。购物中心以"顾客至上"为出发点，以"建设炫亮城市"为目标，打造集购物、休闲、娱乐、饮食、文化及社区服务于一体的大型商业中心，促进城市商业功能完善、人口集聚、消费提升、社会经济发展，塑造充满"惊奇、感动、喜悦"的生活空间，坚持不懈地为当地居民提供更加丰富多彩的生活做出贡献。

PROJECT OVERVIEW

Yantai Development Zone Wanda Plaza is located in the west of Hengshan Road, the north of Shuhe River, the north of Huaihe Road, with a site area of 4.87 hectares and a gross floor area of 130,000 square meters. The shopping center adheres to the principle of "customer first" and takes "constructing a bright city" as the goal. As a big shopping center satisfying the demand of shopping, recreation, entertainment, catering, culture and community service, it promotes the perfection of urban commercial functions, population aggregation, consumption upgrading, social & economical development and the creation of living space filled with "surprise, affection and joy" and constantly contributes to the provision of more colorful life for local citizens.

1 烟台开发区万达广场总平面图
2 烟台开发区万达广场立面图
3 烟台开发区万达广场鸟瞰图

4

5

招商运营

广场市场调研深入，客群剖析精准——立足政策新区，紧抓儿童客群，着力打造"烟西地区"面积大、品类全的"儿童业态"购物中心。招商过程中，充分发挥自有主力店"万达宝贝王"品牌优势，优化垂直动线；二层南区打造5000平方米"潮童区"；三层引进主力店"卡通尼乐园"。

营运为王、创新营销——品牌引进多元化、商户管理标准化。从制定《商户手册》、《营业员手册》，到店长会、晨会逐层宣教，再到开闭店检查、经营检查逐项落实，让顾客感觉到，每个店铺都有特色，每个店铺又"很万达"。品牌活动聚力，万达创新营销——传统与新媒体结合，创新引入"网红营销"，线上与线下互动，打造"6·18"疯狂年中庆。

4 烟台开发区万达广场外立面
5 烟台开发区万达广场2号主入口
6 烟台开发区万达广场正立面
7 烟台开发区万达广场1号主入口

INVESTMENT TENDER & OPERATION

The Plaza has made in-depth market surgery and precise customer analysis - based on the policy of the new area, focusing on the customer group of children and striving to create a large and full-equipped shopping center featured on "children business type" in west Yantai region. In the process of attracting investment, the brand advantage of self-owned anchor store "Wanda Kidsplace" is given full play to, optimizing the vertical circulation; a "Children's Area" of 5,000 square meters is constructed in the southern area on 2F; the anchor store "Cartoony" is located on 3F.

Effective operation and creative marketing - the diversification of brand introduction and the standardization of shop management. From the preparation of Shop Management Manual and Manual for Shop Assistant, to the propaganda in the shop manager meeting and morning meeting, to the checking before opening the store and after closing the store and the operation inspection, customers would feel that each shop has its own characteristic while all shops have the character of Wanda. Brand promotion activities and Wanda creative marketing —- a combination of the traditional and the new media; "on-line celebrity marketing" is introduced to realize online and off-line interaction during the "6.18" Crazy Mid-year Sales.

6

广场外装

广场外装借鉴寓意昌盛繁荣的"万宝盒"形式,立面运用铝板搭配玻璃的现代材料,彰显年轻态的商业气息;以香槟金色为主的色彩搭配,显得亮丽、高贵,犹如坐落于城市中的稀世珍宝。主立面通过三种不同的肌理效果进行渐进组合,显得丰富而有秩序;同时竖向的纹理加强金属质感,使建筑有种"纵横激流千里远,上下厚土万层深"的厚重感。

FACADE DESIGN

The facade design of the Plaza takes the form of the prosperous "treasure box" as a reference. The facade adopts modern materials of aluminum plate and glass to highlight the youthful commercial atmosphere; the color matching is dominated by champion gold, appearing to be bright and noble like a treasure in the city. The main facade adopts progressive combination of three different texture effects, appearing to be lavish and orderly; in the meanwhile, the vertical grain intensifies the metallic feeling, rendering the building with a kind of concordances.

8 烟台开发区万达广场室内步行街
9 烟台开发区万达广场室内步行街
10 烟台开发区万达广场商业落位图
11 烟台开发区万达广场中庭顶棚
12 烟台开发区万达广场中庭

广场内装

烟台开发区万达广场室内设计以白色为基调，运用现代、简约的设计手法，通过局部香槟金色铝板的点缀，辅以有彩色和独立色，增加空间丰富、活跃的感觉；并与外立面遥相呼应，衬托空间的商业文化色彩，突出空间的展示性，使顾客有种"皑如山上雪，皎若云间月"的感受。

INTERIOR DESIGN

The interior design of the Plaza takes white as the tone and adopts modern and simplified design method. Champagne gold aluminum plates are used for embellishment. Painting color and independent colors are also used to increase the richness and vitality of the space; it also echoes with the facade, sets off the commercial and cultural atmosphere of the space and highlights the ornamental value of the space, rendering the customers with a feeling of "as white as the snow on the mountain and as bright as the moonlight shining through the cloud".

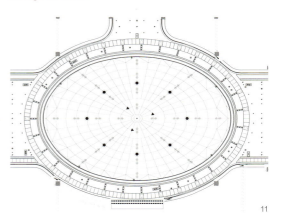

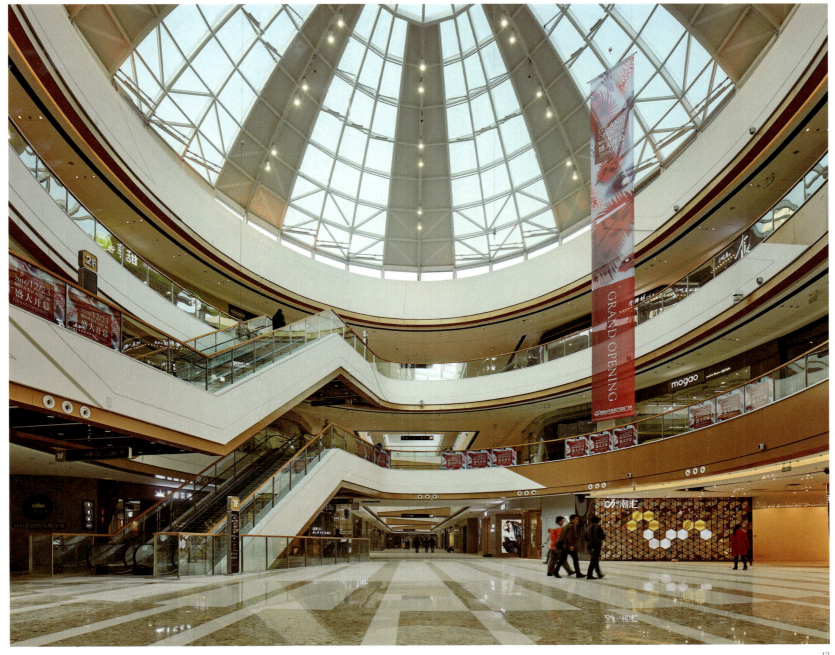

广场景观

广场景观设计主题是"竹韵琴音",隐喻竹子不畏逆境、中通外直、胸怀宽广的优良品格。这是一种取之不尽的精神财富,也是万达广场的精神追求。竹笛演奏的琴声优雅而不失灵动,奔放而不失细腻,是精神文明和物质文明的结合,也与万达追求的经营理念相契合。琴键元素作为主广场的景观主题的"副主题",既增强了广场的韵律感,也体现了万达广场的精神追求,增加景观品质效果。

LANDSCAPE DESIGN

The landscape design of the Plaza adopts the theme of "bamboo & melody", as a metaphor for the excellent qualities of Bamboo - courage, integrity and largeness of mind. This is a kind of inexhaustible spiritual treasure as well as the spiritual pursuit of Wanda Plaza. The sound of bamboo flute performance is elegant, dynamic, rough and smooth. It's a combination of spiritual civilization and material civilization, adhering to the management philosophy of Wanda. As the "subtopic" of the landscape design of the main plaza, the element of "piano-key" not only increases the sense of rhythm for the plaza, but also reflects the spiritual pursuit of Wanda Plaza, increasing the quality and effect of the landscape.

13

13 烟台开发区万达广场绿化
14 烟台开发区万达广场景观
15 烟台开发区万达广场夜景
16 烟台开发区万达广场夜景

14

广场夜景

广场夜景设计元素从万达LOGO中提取素材,将主立面的灯光效果以飘带的方式展现出来。形成一个具有流动性的线性走廊,用灯光来展现立面流动的效果,与广场人流交相呼应。通灯光的流动、人的活动将室外大环境"串联"起来,将商业气息淋漓尽致地展现出来。

NIGHTSCAPE DESIGN

The nightscape design of the Plaza takes Wanda LOGO as a reference, presenting the lighting effect of the main facade in the form of streamers and forming a flowing linear corridor; the lights are used to present the flowing effect of the facade, echoing with the crowds in the square. The outdoor environment is "connected together" by the flowing lights and people's activities, fully revealing the commercial atmosphere.

06

CHAOYANG WANDA PLAZA
朝阳万达广场

OPENED ON : 26th NOVEMBER, 2016
LOCATION : CHAOYANG, LIAONING PROVINCE
LAND AREA : 5.11 HECTARES
FLOOR AREA : 97,000 SQUARE METERS

开业时间：2016 / 11 / 26
开业地点：辽宁省 / 朝阳市
占地面积：5.11 公顷
建筑面积：9.7 万平方米

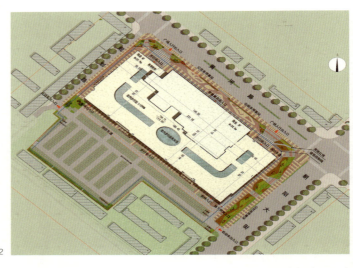

广场概述

朝阳万达广场位于辽宁省朝阳市双塔区朝阳大街及黄河路的交汇处，规划用地5.11公顷，总建筑面积9.7万平方米（地下0.7万平方米，为机电设备用房及卸货区；地上9.0万平方米），地上停车场可容纳630辆。

PROJECT OVERVIEW

Chaoyang Wanda Plaza is located in the junction of Chaoyang Street and Huanghe Road in Shanghai District of Chaoyang City, with a site area of 5.11 hectares and gross floor area of 970,000 square meters (7,000 square meters underground as the room for M&E equipment and loading area; 9,000 square meters above-ground). The ground parking lot can accomodate 630 vehicles.

1 朝阳万达广场外立面
2 朝阳万达广场总平面图

3

招商运营

朝阳万达广场招商策划充分考虑城市周边环境，定位于"城市全客层影响力的时尚购物中心"；核心客群为新组家庭为主，工人阶层、工厂管理阶层为辅；重在引进国际知名品牌、快时尚、儿童业态及特色餐饮连锁品牌及"新奇特"的体验式服务业态，兼顾各层次消费者。

在运营中对招商业态加以调整，不断提升品牌和服务品质，为朝阳人提供休闲娱乐及购物的时尚消费体验。营销活动注重引入当地特色元素，通过场景再现、举办民俗活动及展示当地特色品牌，取得名优产品聚集的效应。

INVESTMENT TENDER & OPERATION

The scheme to attract investment for Chaoyang Wanda Plaza gives full consideration to the surrounding environment of the city, oriented at "a fashionable shopping center in the city with influences of customers of all levels"; the core customer groups are the newly-married couples, supplemented by the working class and people at managerial level in factories; the main investors are intentionally renowned brands, fast fashion, business types for children, chain brands of featured catering and new and unique experience-based service business types, giving consideration to consumers of all levels.

In operation, the business types for attracting investment are adjusted to continuously promote brand and service quality, providing fashionable consumption experience for the leisure, entertainment and shopping of citizens in Chaoyang. The promotion activities lay emphasis on the introduction of local featured elements. An agglomerative effect of quality brands is achieved by reappearance of scenes, the holding of folk activities and the presentation of local featured brands.

4

5

广场外装

建筑立面主体以矩形为元素，通过规律的错动营造出动感、层次丰富的立面效果。门头的设计呼应矩形主题，并强调了主入口。不同颜色铝板或彩釉玻璃的结合使用突出了建筑的商业氛围。设计借鉴佛教七宝琉璃塔中"金"与"玉"的运用，即金色矩形与白色矩形为单元，运用穿插、搭接等建筑设计手法，使之成为城市的"瑰宝"，使广场更具朝阳地域特色。

FACADE DESIGN

The facade of the building takes rectangular as an element. A dynamic and multi-layered facade is created through regular dislocation. The design of door head echoes with the theme of rectangular and highlights the main entrance. The combined utilization of aluminum plates of different Colors or glazing glasses highlights the commercial atmosphere of the building. The design imitates the utilization of "gold" and "jade" in the Shippo Glazed Tile Pagoda in Buddhism. That is, gold rectangular and white rectangular are taken as units, and the units are interweaved and lap-jointed in the design, making the Plaza a "treasure" of the city and rendering it with more re-gional characteristics of Chaoyang.

3 朝阳万达广场外立面
4 朝阳万达广场外立面
5 朝阳万达广场外立面

6 朝阳万达广场室内步行街
7 朝阳万达广场中庭
8 朝阳万达广场室内步行街电梯
9 朝阳万达广场商业落位图

广场内装

朝阳万达广场室内空间设计以蜿蜒的折线为主要装饰元素。金色金属质感的材质从明亮大方的白色主体中"跳出"，现代又不失自然协调；同时作为外立面金色元素在室内的延续，不仅点缀室内商业氛围，而且呼应了整体设计格调。合理的材质选择，塑造了美观大方、整体感强的商业环境氛围，突出了空间的纯粹与灵动，实现了以"肌理的融合和对比"为概念对商业空间进行精心刻画的初衷。

INTERIOR DESIGN

The interior design of Chaoyang Wanda Plaza takes winding lines as the main decorative elements. The gold material of metallic texture "jumps out" from the bright and generous white body, which looks modern, natural and coordinative; in the meanwhile, as an indoor continuance of the gold elements on the facade, it not only adorns the indoor business atmosphere, but also echoes the overall design style. The reasonable selection of materials creates a beautiful, generous and integrated commercial environment, highlighting the purity and mobility of the space and realizing the original design purpose, which is to create an elaborate commercial atmosphere under the concept of "the fashion and contrast of texture".

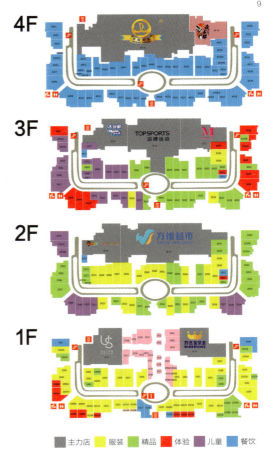

10

10 朝阳万达广场夜景
11 朝阳万达广场夜景
12 朝阳万达广场夜景

11

12

广场夜景

灯光设计风格以"打造大美朝阳"的环境形象为目标，通过变幻多彩的灯光系统与建筑立面幕墙的完美结合，在映衬建筑秀美的夜景体态的同时，与景观设计理念相结合，提升了广场的整体夜景效果；同时将朝阳的夜间形象更好地表现出来，丰富朝阳人民的夜生活。

NIGHTSCAPE DESIGN

The lighting design sets "creating the beautiful image of Chaoyang" as a goal. Through the perfect combination of lighting system and building facade, the lighting design not only sets off the beautify nightcap of the building, but also combines with the landscape design concept, promoting the overall nightscape effect of the Plaza; meanwhile, it makes the night image of Chaoyang more prominent and enriches the night life of the citizens of Chaoyang.

07
BOZHOU WANDA PLAZA
亳州万达广场

OPENED ON : 12th AUGUST, 2016
LOCATION : BOZHOU, ANHUI PROVINCE
LAND AREA : 38,800 SQUARE METERS
FLOOR AREA : 147,500 SQUARE METERS

开业时间： 2016 / 8 / 12
开业地点： 安徽省 / 亳州市
占地面积： 3.88 万平方米
建筑面积： 14.75 万平方米

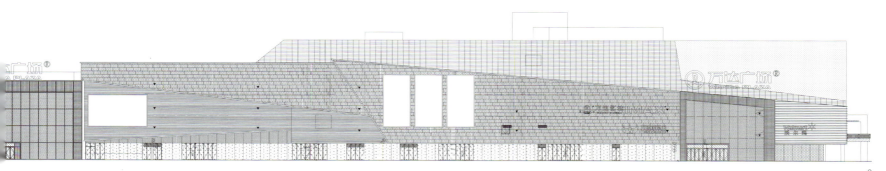

广场概述

亳州万达广场位于安徽省亳州市南部新区核心位置，处于希夷大道东侧，杜仲路南侧，由商业综合体、写字楼、商务公寓、室外步行街、特色商业街及住宅等业态组成，集文化艺术、旅游休闲、商业娱乐、商务办公、高品质居住"五大功能"于一体，为亳州市南部片区高品质精品项目。

PROJECT OVERVIEW

Bozhou Wanda Plaza is located in the core of the southern new district of Bozhou City, Anhui Province, to the east of Xiyi Avenue and the south of Duzhong Road. It consists of various business types, including commercial complex, office buildings, commercial apartments, outdoor pedestrian streets, featured commercial streets and residential building. Being a high-quality project in the southern area of Bozhou, it provides "five major functions" of culture & art, tourism recreation, commercial entertainment, commercial office and high-quality living in one spot.

1 亳州万达广场外立面
2 亳州万达广场立面图
3 亳州万达广场总平面图

广场外装

亳州万达广场外装立面契合城市的文化活力主题，颇具"玄鸟飞腾"之魄力，"武术拳法"之动感。其整体线条的设计使大型体块富有动态之美；其整体颜色配置丰富中又不乏稳重之感。体块状的表面被分割为倾斜和折叠的"碎片"使建筑不再单调无趣。通过对表面"碎片"深度的有机变化组合产生丰富多变的阴影效果，给城市增添了变化的活力。

FACADE DESIGN

The facade design of Bozhou Wanda Plaza agrees with the theme of dynamic culture of the city, presenting the vigor of "flying birds" and the dynamics of "martial arts". The design of overall lines renders the big massing with the beauty of mobility; the overall color configuration is both rich and sedate. The blocky surface of the facade is cut into oblique and folded "debris", making the building more interesting. A diversified shadow effect is created via the organic changes and combinations of the "debris" on the surface, adding the vitality of change to the city.

4 亳州万达广场外立面
5 亳州万达广场主入口
6 亳州万达广场室内步行街
7 亳州万达广场中庭

广场内装

亳州万达广场依托地域背景"药都"之誉,以"购物"与"购药"为共性的概念出发,借用药材采集的原始工具"背篓"形态作为表现切入,抽取"背篓"编织构成精髓,进行概念重组、变化,让购物广场在现代气息与地域色彩中交织绽放。编织艺术作为原始形态,其流畅的线条与粗犷的肌理结合,装饰以多种材质、灯光的现代表现,使广场的立面形态活泼多姿,给商业空间注入自然的人文气息,提供轻松、愉悦的购物环境。

INTERIOR DESIGN

With the regional background of Bozhou as "the city of medicine" and based on the concept of the generality of "shopping" and "buying medicine", the design of Bozhou Wanda Plaza takes "pack basket", the original tool for medicine collecting, as the form for presentation and absorbs the essential knitted structure of "pack basket". With recombination and change of concepts, the shopping Plaza blooms in the interweaved atmosphere of modern style and regional characteristic. Knit art being taken as the primitive form, its smooth lines and straightforward combination of texture, with the decoration of modern presentation of various materials and lights, makes the facade of the Plaza lively and colorful, brings the breath of natural humanity into the commercial space and provides a delightful and relaxing shopping environment.

8

广场景观

广场景观设计理念为"涡河画卷"——将"涡河文化"加以艺术化处理，以画卷形式展示亳州的历史文明。设计以亳州"历史、人文、物产"入题，描绘亳州由古代走向未来的发展进程，绘就一幅亳州景观的雄伟画卷。广场景观设计中突出运用"芍药花"——亳州素有"中华药都"美誉；"芍药花"既是"亳药"精粹，也是亳州的市花。北广场以"亳芍精粹"为景观立意，以"芍药花"花瓣形成的曲线叠水结合互动空间将主广场景观引入高潮。

LANDSCAPE DESIGN

The landscape design concept of the Plaza is "the Scroll Painting of the Guohe River" - with artistic processing of the "Culture of the Guohe River", the history and civilization of Bozhou is presented in the form of a scroll painting. The design takes "history, humanity and property" as the theme to describe the development process of Bozhou from the past to the future, drawing a magnificent picture of the landscape of Bozhou. The landscape design of the Plaza highlighted the element of "peony flower" - Bozhou has always been praised as "City of Medicine in China"; "peony flower" is not only the essence of "Bozhou Medicine", but also the city flower of Bozhou. The design concept of the north square is the "essence of Bozhou peony flower", pushing the landscape in the main square to the climax with the curving cascading waterfalls in the shape of the petal of "peony flower" and interaction space.

8 亳州万达广场景观
9 亳州万达广场路灯
10 亳州万达广场室外步行街
11 亳州万达广场室外步行街

9

室外步行街

亳州万达广场的地块区位、性质等特征决定了它将超越普通的商业功能，从而担负展示城市性格、提升城市品牌的使命。亳州万达金街以"新中式"风格为主题，既体现中式的典雅，又展现当代设计简洁明快的风范。步行街体现了规划的"均好性"与平面布局的"实用性"，为商户和顾客创造了价值。

OUTDOOR PEDESTRIAN STREET

The location and characteristics of Bozhou Wanda Plaza determines that it would go beyond the identity of an ordinary commercial building, and should carry the mission of displaying the character of the city and enhancing the brand of the city. t Taking "new-Chinese" style as the theme, Bozhou Wanda Gold Street not only shows the Chinese-style elegance, but also embodies the features of simple and elegant style of modern design. The pedestrian street reflects the "balance" of the planning and the "practicability" of the layout, creating value for the merchants and the customers.

08

JIXI WANDA PLAZA
鸡西万达广场

OPENED ON : 15th JULY, 2016
LOCATION : JIXI, HEILONGJIANG PROVINCE
LAND AREA : 5.84 HECTARES
FLOOR AREA : 399,500 SQUARE METERS

开业时间： 2016 / 07 / 15
开业地点： 黑龙江省 / 鸡西市
占地面积： 5.84 公顷
建筑面积： 39.95 万平方米

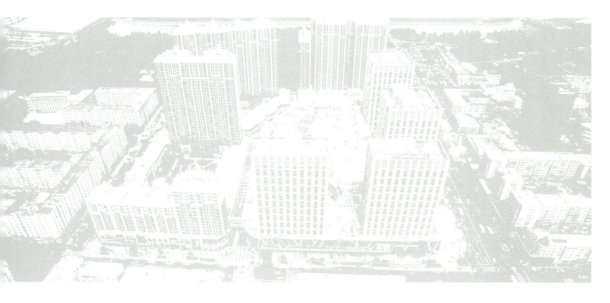

1 鸡西万达广场鸟瞰图
2 鸡西万达广场总平面图

广场概况

鸡西万达广场坐落在鸡西市中心大街以西、祥光路以北，占地面积5.84公顷，总建筑面积39.95万平方米。其中地上建筑面积32.15万平方米，地下建筑面积7.8万平方米。广场由五部分组成：大商业、公寓楼、住宅区、回迁住宅商业及室外商业街。其中大商业建筑面积12.36万平方米，销售物业建筑面积21.82万平方米，回迁住宅及商业建筑面积5.60万平方米，配套物业0.17万平方米。

招商运营

鸡西万达广场是"吃喝玩乐购"的"一站式"购物中心，拥有多个首次登陆鸡西的品牌，引领当地商业潮流。开业共计168个品牌，包括万达影城、宝贝王、物尔美超市、苏宁易购、大玩家等主力店，以及国际连锁餐饮麦当劳、肯德基、必胜客等。

广场开业庆典活动举办了"海底萌宠"及百台机器人的机械舞表演，体验高科技为生活带来的便利及惊喜；配合"海洋摇滚音乐节"打造一个"鸡西不夜城"。活动期间邀请黑龙江省知名主持人参与活动，全城造势，现场与顾客进行游戏互动，劲歌热舞 轰动全城。

PROJECT OVERVIEW

Jixi Wanda Plaza is located to the west of the Central Avenue and to the north of Xiangguang Road in Jixi City, with a site area of 5.84 hectares and a gross floor area of 399,500 square meters (321,500 square meterss above-ground and 78,000 square meters underground). The Plaza consists of five parts: commercial building, apartment, residential area, residential & commercial building and outdoor commercial area; the big commercial building covers an area of 123,600 square meters; the sales property 218,200 square meters; the residential & commercial building 56,000 square meters; the support property 1,700 square meters.

INVESTMENT TENDER & OPERATION

Jixi Wanda Plaza is an "one-stop" shopping center that satisfies the customers' desire to "eat, drink, play, entertain and go shopping". It has several brands that are first introduced to Jixi, leading the local business trend. At the opening, there are 168 brands all together, including anchor stores (i.e. Wanda Cinema, Wanda Kidsplace, Wu-mark, Suning and Big Player) and international chain catering stores (i.e. McDonald's, KFC and Pizza hut).

During the opening ceremony, exhibition of "Adorable Pets under Sea" and poppin performance with hundreds of Robots are to held to give the customers a chance to experience the convenience and surprise that high-technology brings to life; Jixi is built into a "sleepless city" in corporation with the "Ocean Rock Music Festival". During the event, a famous host in Heilongjiang Province is invited to activate the audiences, and the audiences are invited to play games, singing and dancing. The event is phenomenal in the city.

3

广场外装

广场外装设计提取"白桦林"为元素，使建筑立面线条苍劲挺拔，宛若少女亭亭玉立；色彩上运用具有地方特色的秋季"金色"，似秋天孕育的丰收与喜悦。大商业立面设计采用现代的风格，将弧形线条和垂直线优雅地相接，散发简约、轻盈的愉悦感。

FACADE DESIGN

The facade design of the Plaza adopts the element of birch forest, making the lines of the building facade hardy and straight like a slim and graceful maid; as for colors, the local color - "golden in the fall" is applied as a metaphor of the harvest and joy of the fall. The facade of big commercial area adopts modern style design, and the arc-shaped lines and perpendicular lines are connected elegantly to radiate a simple and delighted sense of joy.

3 鸡西万达广场主入口
4 鸡西万达广场立面图
5 鸡西万达广场外立面

6 鸡西万达广场中庭
7 鸡西万达广场室内步行街
8 鸡西万达广场室外步行街
9 鸡西万达广场室外步行街

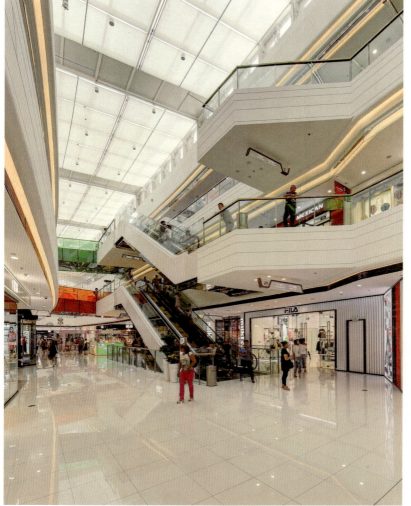

广场内装

广场内装设计理念为"秋天白桦林",力求避免过度的装饰,力争在空间设计上突破创新。地面及连桥"白桦林"元素贯穿整个室内步行街。连桥的白桦林图案玻璃及彩色地砖相互呼应,使人如置身于白桦林中,感受秋天的静谧美好。

INTERIOR DESIGN

The interior design concept of the Plaza is "birch forest in fall", striving to avoid excessive decoration and get a breakthrough in the spatial design. The element of "birch forest" on the floor and the gallery bridge runs through the whole interior pedestrian street. The pattern of birch forest on the glasses in the bridge echoes with the colored floor tiles, making the customers feel like they are actually in the birch forest and enabling them to enjoy the beauty and quietness of fall.

室外步行街

鸡西万达室外街是具有当地文化气息的商业步行街，缔造"文商并重，古今融合"的商业气息。通过一系列的特色小品，既塑造室外步行街的文化调性，贴近当地文化，具有亲切感和归属感，又聚拢人气，为当地民众提供休闲娱乐的好场所。

OUTDOOR PEDESTRIAN STREET

The outdoor pedestrian street of Jixi Wanda Plaza is a commercial pedestrian street with local cultural atmosphere, radiating the commercial atmosphere of "attaching equal importance to culture and commercial and integrating the past with the present". The cultural tone of the outdoor pedestrian street is created through a series of featured sketches. It not only accords with local culture, rendering a sense of closeness and belonging, but also reassembles the throngs, becoming a good place for the recreation and entertainment of local citizens.

09

MUDANJIANG WANDA PLAZA
牡丹江万达广场

OPENED ON: 22nd JULY, 2016
LOCATION: MUDANJIANG, HEILONGJIANG PROVINCE
LAND AREA: 3.83 HECTARES
FLOOR AREA: 183,700 SQUARE METERS

开业时间: 2016 / 07 / 22
开业地点: 黑龙江省 / 牡丹江市
占地面积: 3.83 公顷
建筑面积: 18.37 万平方米

广场概述

牡丹江万达广场位于牡丹江市西安区，用地面积3.83公顷，总建筑面积18.37万平方米，其中购物中心建筑面积为13.53万平方米（含地上8.13万平方米和地下5.4万平方米）。广场将大型购物中心和文化旅游特色街区有机结合，以"一站式"体验满足购物、休闲、娱乐和餐饮等需求，包括购物中心、精装公寓、万达华府、万达金铺等多种业态，形成和谐共生的城市中心顶级资源聚合体，全面带动牡丹江的城市发展和商业升级。

招商运营

牡丹江万达广场于2016年7月22日盛大开业，作为综合性"一站式"购物中心，拥有175个品牌，分为餐饮类、服装类、精品类和体验类；主力店包括万达影城、宝贝王、大玩家等品牌，以及世界知名连锁餐饮品牌。

广场通过举办各种活动，推动全面联动、深度合作，以体验业态为主，带动服装、餐饮业态，以提升经营业绩、拉动家庭客流为导向，推出火爆全城的商品优惠活动，实现客流销售双提升，轰动牡丹江全市，宾朋满店。

PROJECT OVERVIEW

Mudanjiang Wanda Plaza is located in Xi'an District of Mudanjiang City, with a site area of 3.83 hectares and a gross floor area of 183,700 square meters; the shopping mall covers an area of 135,300 square meters (81,300 square meters above-ground and 54,000 square meters underground). The Plaza organically combines the function of a large shopping center and a featured block for cultural tourism. It satisfies the needs for shopping, recreation, entertainment and catering with one-stop experience and covers various business types of shopping center, fine-decorated apartment, Wanda Palace and Wanda Golden Store. It serves as a harmonious top resources integration in city center, comprehensively bringing along the city development and commercial upgrading of Mudanjiang.

INVESTMENT TENDER & OPERATION

Mudanjiang Wanda Plaza opened on July 22nd, 2016. As an comprehensive "one-stop" shopping center, it has 175 brands under the categories of catering, clothing, boutique and experience; the anchor stores include Wanda Cinema, Wanda Kidsplace and Big Player and international chain catering brands.

Through various activities, the Plaza advocates overall unification and in-depth cooperation. Most attention is given to the experience business type to promote the sales of the business type of close and catering. Aiming at promoting operational performance and increasing family customers, popular sales and promotion activities are pushed forward to realize the dual progress of client flow and sales. The sensational activities attract tones of customers in Mudanjiang City.

1 牡丹江万达广场鸟瞰图
2 牡丹江万达广场总平面图

广场外装

牡丹江有着全国降雪量最大的大林海,被世人赞誉为"中国雪乡"。"林海雪原"成为当地属性之一。设计充分挖掘当地地域特色,从"林海雪原"中吸取设计元素——"松林"与"雪海"成为其中主要符号元素——立面以硬朗的折线化为山脊,穿插交织的体块构成山峦走势;形体间错落的层次关系仿佛山间苍松劲柏,又好似层层的山林和层层的雪山相映成趣,使广场具有强烈的体量感和地域属性。

FACADE DESIGN

Mudanjiang is the home to the large forest with the heaviest snowfall in China, and it's lauded as the "Snow City of China". "Snowy forest" has become one of the local attribute. The design makes full use of local and regional features and absorbs design elements from "snowy forest" - "pine forest" and "intense snow" become the major symbol elements - the facade takes straight broken lines as metaphors for mountain ridges, and the interweaved masses form the direction of the mountains; the hierarchical and staggered masses look like the hardy pines and cypresses in the mountains as well as layers of forests and snowy mountains contrasting finely with each other, rendering the Plaza with intense sense of volume and regional attributes.

3 牡丹江万达广场外立面
4 牡丹江万达广场外立面
5 牡丹江万达广场外立面

广场内装

牡丹江万达广场内装设计抽取了雪域特色元素，将"林海雪原"这一主题抽象化。整个商场空间呈现出层层叠叠的积雪线条，通过简洁明快的线条表现和轻盈的色彩搭配，烘托商场空间氛围，形成与"雪城"牡丹江相协调的特有商业氛围。

INTERIOR DESIGN

The interior design of Mudanjiang Wanda Plaza extracts the featured element of snowy area and abstracts the theme of "snowy forest". The whole shopping mall is filled with stacked and layered snow lines. The atmosphere of the shopping mall is set off by the concise and lively lines and lightsome color matching, forming an unique commercial atmosphere that agrees with the "snow city" Mudanjiang.

6 牡丹江万达广场室内步行街
7 牡丹江万达广场椭圆中庭
8 牡丹江万达广场圆中庭
9 牡丹江万达广场室内步行街

10
URUMQI JINGKAI WANDA PLAZA
乌鲁木齐经开万达广场

OPENED ON: 12th AUGUST, 2016
LOCATION: URUMQI, XINJIANG UYGUR AUTONOMOUS REGION
LAND AREA: 4.56 HECTARES
FLOOR AREA: 173,600 SQUARE METERS

开业时间：2016 / 08 / 12
开业地点：新疆维吾尔自治区 / 乌鲁木齐市
占地面积：4.56 公顷
建筑面积：17.36 万平方米

广场概述

乌鲁木齐经开万达广场位于乌鲁木齐经济技术开发区，处于阿里山街以南、万寿山街以北，东临玄武湖路，地块西侧为金街及销售住宅区，占地面积4.56公顷，建筑面积17.36万平方米，其中大商业建筑面积15.41万平方米。广场由购物中心、室外步行街和三栋SOHO公寓组成，主要业态为室内步行街及主力店和次主力店。

运营招商

乌鲁木齐经开万达广场是全方位、全覆盖的新型购物中心，购物中心室内步行街长320米，涵盖国际时尚精品、潮流时尚服饰、数码生活体验、儿童零售和中西特色餐饮等精品特色业态。广场拥有亚洲排名第一的院线——万达影城，可以提供极致观影体验的IMAX影厅；拥有专为中国家庭打造的创新型动漫亲子乐园"万达宝贝王"。

乌鲁木齐经开万达广场经常举办地方特色鲜明的活动，如"新疆首届经典旅游商品美食展"。"美食展"集纳了新疆各类特色商品、精品美食，展现了新疆旅游商品的品质；现场设大型主题展位60个，面积近1万平方米，成为展示"大美新疆"的重要平台。

PROJECT OVERVIEW

Urumqi Jingkai Wanda Plaza is located in the Economic Development Area of Urumqi, south to Alishan Street, north to Wanshoushan Street, west to Xuanwu Lake Road and east to Gold Street and commercial residential area, with a site area of 4.56 hectares and a gross floor area of 173,600 square meters; the large commercial building covers an area of 154,100 square meters. The Plaza consists of a shopping center, outdoor pedestrian streets, three SOHO apartments, with the main business type of indoor pedestrian street, anchor stores and sub-anchor stores.

INVESTMENT TENDER & OPERATION

Urumqi Jingkai Wanda Plaza is an all-around new shopping mall with full coverage with a 300m indoor pedestrian street, covering featured business types of international fashion boutiques, trendy and fashionable cloth, digital life experience, retail for children, Chinese and Western catering, etc.. The Plaza has the Top 1 Cinema in Asian - Wanda Cinema with IMAX movie hall that provides the ultimate viewing experience; it also has "Wanda Kidsplace", a creative cartoon parent-child paradise specially created for Chinese families.

Events with distinctive local features are held in the Urumqi Jingkai Wanda Plaza, for example the "First Classic Tourism Commodities and Food Fair in Xinjiang Province". The food fair collects various featured commodities and quality food in Xinjiang and shows the quality of Xinjiang's tourism commodities; there are about 60 large themed booths on the site, covering an area of about 10,000 square meters, which becomes an important platform to show the "Grand Beauty of Xinjiang".

1

PART C　WANDA PLAZA　万达广场

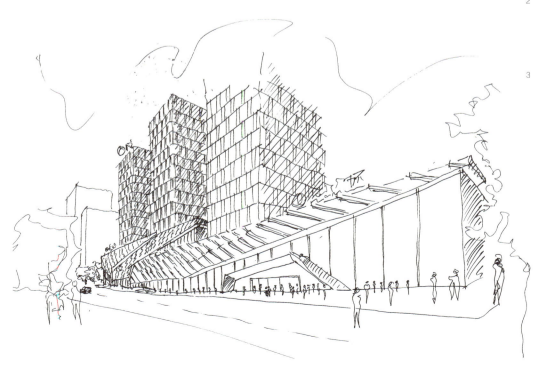

1 乌鲁木齐经开万达广场总平面图
2 乌鲁木齐经开万达广场外立面
3 乌鲁木齐经开万达广场概念草图

4
4 乌鲁木齐经开万达广场3号主入口
5 乌鲁木齐经开万达广场主入口
6 乌鲁木齐经开万达广场室内步行街
7 乌鲁木齐经开万达广场中庭

广场外装

乌鲁木齐是连接新疆与内地的交通、通信总枢纽，也是丝绸之路的重要节点。万达广场融入其中并作为城市新的地标，充分展现城市新风貌。广场外装以"薄纱"作为设计概念，在东侧主立面上采用白色、深灰色和浅蓝色三种颜色的铝板，运用富于动感的线条，通过几何形态的拼接形成轻盈简洁的造型。此外，投射在建筑上的几何状阴影，也赋予建筑立面以和谐的对比，形式丰富的"肌理感"和"层次感"。

FACADE DESIGN

Urumqi is the main traffic and communication hub connecting Xinjiang with the mainland and an important node in the Silk Road. As a new landmark of the city, Wanda Plaza is integrated into the city and fully reveals the new look of the city. The exterior design of the Plaza takes "veiling" as the concept. The main facade on the east side adopts white, dark grey and light blue aluminum plates, and forms a lightsome and concise form with dynamic lines and the splicing of geometric patterns. In addition, the geometric shadow projected on the building also renders a harmonious contrast to the building facade, which is rich in "texture sense" and "sense of layer".

5

广场内装

乌鲁木齐经开万达广场以"丝绸之路"作为设计概念。设计提取"丝绸之路"地域显著人文、地理元素——沙漠、绿洲、城堡等融入内装造型,以此体现典型的自然风貌与城市(城堡)建筑的特性,展望"一带一路"的美好前景。室内造型将设计理念加以确实地贯彻——圆中庭侧板犹如金色沙漠,皱褶如沙浪连绵不绝,连桥侧板体现城市(城堡)建筑的造型特性,直街侧板错落如沙浪翻滚……"丝绸之路"的气息无处不在。

INTERIOR DESIGN

Urumqi Jingkai Wanda Plaza takes "Silk Road" as the design concept. The design extracts prominent cultural and geographical elements from the region of "Silk Road" - desert, oasis and castle and merges them into the interior design, so as to embody the typical natural landscape and the characteristics of city (castle) buildings and look ahead to the great prospect of the "One Belt One Road" policy. The interior modeling put the design concept into actual practice - the side plate of the round atrium looks like a golden desert with folds like rolling and endless sand waves; the connecting bridge side plate embodies the modeling feature of city (castle) buildings; and the straight street side plate is staggered like the rolling sand wave... The small of "Silk Road" is everywhere.

11
DEYANG WANDA PLAZA
德阳万达广场

OPENED ON: 17th SEPTEMBER, 2016
LOCATION: DEYANG, SICHUAN PROVINCE
LAND AREA: 3.85 HECTARES
FLOOR AREA: 129,800 SQUARE METERS

开业时间： 2016 / 09 / 17
开业地点： 四川省 / 德阳市
占地面积： 3.85 公顷
建筑面积： 12.98 万平方米

广场概述

德阳万达广场位于德阳市旌阳区黄河新城,是集大型休闲购物中心、SOHO公寓、步行商业街及高端住宅等多种业态于一体的综合体,也是德阳目前规模与设施领先的"一站式"购物中心。广场能满足休闲、娱乐、餐饮和购物等多项需求,提供全程体验消费模式,适合不同年龄层次的消费者。广场将辐射德阳及周边消费人群,是德阳城市新地标、城市商业休闲中心和"一站式"高端生活区。

PROJECT OVERVIEW

Deyang Wanda Plaza is located in Huanghe New Town in Jinyang District of Deyang. It's a complex combining various business types of large recreational shopping mall, SOHO apartment, pedestrian commercial street and high-end residential buildings. It's also a large-scale one-stop shopping center with advanced facilities in Deyang. The Plaza can satisfy multiple needs of recreation, entertainment, catering and shopping, provides experience-type consumption model and is suitable for customers of all ages. The Plaza will attract customers in Deyang and surrounding areas and become the new landmark, urban commercial and recreational center and one-stop high-end living area in Deyang New Town.

1 德阳万达广场鸟瞰图
2 德阳万达广场总平面图

3

3 德阳万达广场外立面
4 德阳万达广场主入口
5 德阳万达广场中庭

4

广场外装

设计紧扣"发展"之主旨，建筑形态中间厚重、两端轻盈，如展翅飞翔的鸿鹄，象征腾飞的德阳。立面采用标准化单元构件形成幕墙表皮。其材料以银白色菱形铝板为主，通过单块铝板的互相咬合形成"鳞片"式的拼接纹理，在局部反向扭转"鳞片"，形成丰富的建筑表皮效果。彩釉玻璃与金属形成的金银两色，既创造了浓厚的商业氛围，又彰显面向未来的时代精神。

FACADE DESIGN

The design sticks to the theme of "development". The building is dense in the middle and lightsome on two sides which looks like a large swan with its wings stretched as a metaphor for the souring Deyang. The facade adopts standard unit members to form the surface of the curtain wall. The material is mainly silver-white diamond aluminum plate. The aluminum plates interlock with each other to form a "scale-like" splicing texture. In some parts, the "scales" are reversed to enrich the effect of the building facade. The gold and silver glazing glass and metal not only create a strong commercial atmosphere, but also manifests a future-oriented spirit of the time.

广场内装

德阳万达广场以建筑外观"鳞韵"主题，通过简化提炼为斜线组合获得设计灵感。斜线作为建筑外部的主元素与室内设计概念的来源，成为连接整个项目的纽带。室内各区的斜面造型，运用不同材质混搭组合，通过虚实对比的手法，呈现出简练造型中孕育丰富细节变化的视觉效果。其中，步行街及椭圆中庭通过错层的斜线造型，搭配亮面不锈钢条状造型，以渐变的形式散布在洁白的面板上，呈现出跳跃的视觉效果，营造出活泼温馨的购物氛围。

INTERIOR DESIGN

The exterior design of Deyang Wanda Plaza takes "Scale-like" as the theme and is inspired by the combination of simplified oblique lines. As the main element of building facade and the source of interior design concept, the oblique line becomes the link of the whole project. Through the mix and combination of different materials and the method of void-solid comparison, the modeling of oblique line in different interior areas presents diversified and lavish visual effects in the concise modeling. The staggered modeling of oblique line in the pedestrian street and oval atrium, matched with stainless steel bars, is scattered on the white plate in a gradually varied form to present a jumpy visual effect and create a lively and cozy shopping atmosphere.

WANDA
COMMERC
PLANN

01

WANDA REIGN SHANGHAI
上海万达瑞华酒店

OPENED ON : 18th JUNE, 2016
LOCATION : HUANGPU DISTRICT, SHANGHAI
LAND AREA : 36,400 SQUARE METERS

开业时间： 2016 / 06 / 18
开业地点： 上海市 / 黄埔区
建筑面积： 3.64 万平方米

酒店概况

上海万达瑞华酒店是七星级超豪华酒店，坐落于闻名遐迩的上海外滩，紧邻黄浦江，酒店正对上海外滩著名的十六铺码头，可同时拥有外滩、黄浦江及浦东天际线等上海最经典的繁华美景。酒店距上海浦东国际机场42公里、上海虹桥国际机场16公里，至上海火车站仅6.4公里。酒店拥有193间高贵典雅的客房（含14间套房），套内面积45~288平方米，设有五家风格迥异的餐厅和酒吧，为客人提供多元化的餐饮选择，包括由法国传奇米其林星级大厨马克·曼努主理的"瑞酷餐厅"、全日餐厅"美食汇"、"游宴一品"淮扬餐厅、日式料理及大堂酒廊。位于酒店3层的720平方米无柱式大宴会厅和3间多功能会议室，适合举办各种庆典、晚宴、私人聚会和商务活动。

PROJECT OVERVIEW

Wanda Reign Shanghai is a seven-star deluxe hotel. It is located at the famous Shanghai Bund, and is close to the Huangpu River, It overlooks the Bund, Huangpu River, Pudong skyline and other splendid downtown views. In front of the hotel is the famed the Shiliupu Dock(Sixteen Shop Wharf). It is 42 km from Shanghai Pudong International Airport, 16 km from Shanghai Hongqiao Airport, and only 6.4 km from Shanghai Railway Station. The hotel has 193 noble and elegant guest rooms (including 14 suites) with net floor areas of 45~288 square meters. Five restaurants and bars with different styles are available to offer diversified food, including "RUI KU" restaurant managed by the French chef and Michelin star winner Marc Meneau, the 24-hour restaurant "Café Reign", "RIVER DRUNK" presents high quality grand Huaiyang cuisine specializing in fresh seafood, a Japanese food restaurant and a lobby bar. The 720 square meters pillarless banquet hall and 3 multi-function meeting rooms on 3F are suitable for celebrations, dinners, parties and business events.

PART D　WANDA HOTEL
万达酒店

1 上海万达瑞华酒店总平面图
2 上海万达瑞华酒店外立面

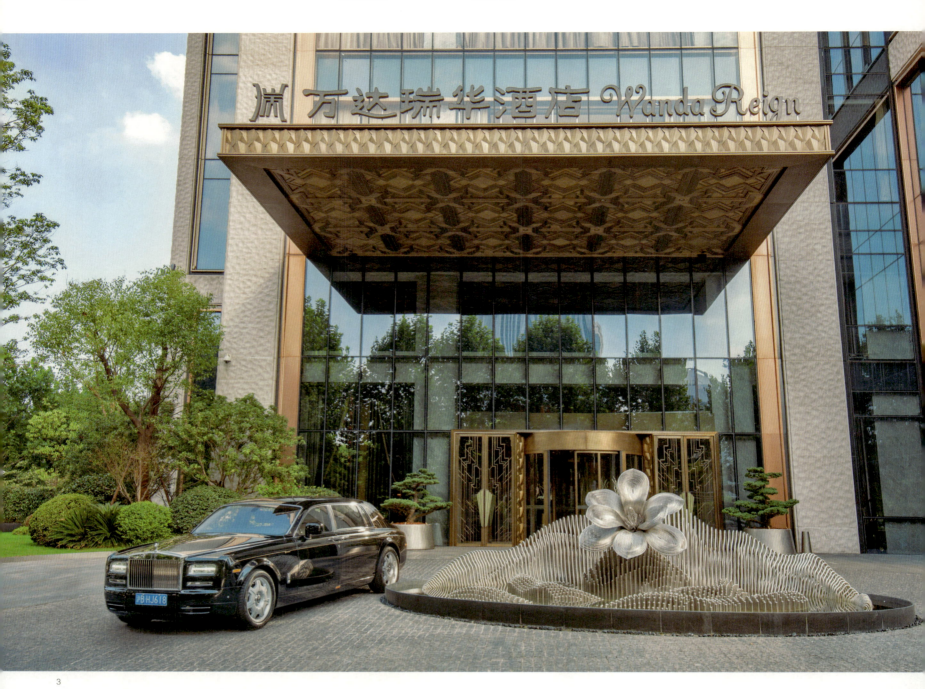

3 上海万达瑞华酒店入口
4 上海万达瑞华酒店外立面

酒店外装

上海瑞华酒店外立面由世界顶级建筑大师、普利策奖得主Forster+Partner事务所精心打造，融合了"古典外滩"和"老上海城区"两种城市肌理——通体玻璃的柱形建筑依照不同高度呈阶梯排布，如同黄浦江边跳跃起伏的音律，形成了美丽的都市天际线；金属线条装饰使得玻璃建筑形体更加挺拔和精致，玻璃上显现出周边黄浦江景的倒影，使得上海瑞华酒店分外妖娆多姿。

FACADE DESIGN

Elaborately designed by the world's top architects and Pulitzer Prize winner Forster+Partner firm, the facade of Shanghai Reign Hotel integrates two different kinds of city textures - "Classic Bund" and "Old Shanghai Urban Area". The column-shaped building fully made with glass is arranged like a flight of stairs, just like the melody at the Huangpu River, and forms a beautiful urban skyline. The metallic lines make the glass building tall and delicate. The beautiful scene along the Huangpu River is reflected on the glass, adding charms to the Reign Hotel.

5

6

5 上海万达瑞华酒店外立面
6 上海万达瑞华酒店外立面
7 上海万达瑞华酒店景观
8 上海万达瑞华酒店绿化
9 上海万达瑞华酒店花坛
10 上海万达瑞华酒店水景

酒店景观

以外滩历史建筑为代表的"传统性"与都市商业区需求的"现代性"要素相融合，形成多功能化的商业酒店建筑。整个景观设计采取"珠玉圆润"的形态并转化为景观元素；同时，铺装设计力图反映建筑立面的线性元素，形成硬朗稳重的整体风格，使环境与建筑珠联璧合。铺装参考了日出到日落的光线变化而做出相应设计，在屋顶花园也使用了铜镜作为设计元素，利用光的折射更强化了设计效果。

LANDSCAPE DESIGN

"Traditionality" represented by the old buildings at the Bund and the "modernity" required by an urban commercial area are combined to shape the multi-functional commercial hotel building. The overall landscape is designed into a round bead, which is turned into an element of the landscape. Meanwhile, the paving is so designed to highlight the lines at the facade so as to achieve tough, firm overall style - thus, the environment and the building are put together in harmony. The paving is designed with reference to the change of beams from sunrise to sunset. As an element of design, the copper mirror at the roof garden utilizes light reflection to highlight the design effects.

02
WANDA VISTA ZHENGZHOU
郑州万达文华酒店

OPENED ON : 25th MARCH, 2016
LOCATION : ZHENGZHOU, HE'NAN PROVINCE
LAND AREA : 47,400 SQUARE METERS

开业时间：2016 / 03 / 25
开业地点：河南省 / 郑州市
建筑面积：4.74 万平方米

酒店概况

郑州万达文华酒店为六星豪华酒店，设在位于郑州繁华商业区的金水万达中心三十三至五十层，有292间典雅舒适、可俯览绿城郑州美景的豪华客房与套房。酒店内设有全日餐厅、中餐厅和大堂酒廊。酒店三层有1400平方米的无柱式大宴会厅，层高10米，配备100平方米高清LED屏幕；另有7间风格各异的多功能厅。酒店设施齐备，设有室内恒温游泳池、健身房、水疗中心、桑拿房和瑜伽房。

酒店外装

郑州万达文华酒店设计方面注重融合大宋王朝古都风韵与现代都市优雅风尚为一体，加以万达文华酒店个性、精致、愉悦的品牌特征，在设计中充分展示了无处不在的千年古都风采。设计将"郑汴古都遗风"与现代都市风格相结合进行创作，酒店随处可见古色古香之设计细节，使宾客在酒店中时刻感受到千年古都的优雅与时尚。

PROJECT OVERVIEW

Wanda Vista Zhengzhou is a six-star hotel that occupies 33F~50F of Jinshui Wanda Center in the busy commercial district of Zhengzhou. It has 292 elegant, comfortable, luxury guest rooms and suites that overlook the beautiful landscape of the green city. There are all day dining, Chinese restaurant and lobby bar in the hotel. On 3F there is a 1400 square meters pillarless banquet room with a 10m floor height and a 100 square meters HD LED screen equipped. There are also 7 multi-purpose rooms with different design styles. The hotel is well-equipped with indoor swimming pool, gym, spa center, sauna room and yoga room.

FACADE DESIGN

The design of Wanda Vista Zhengzhou combines the ancient style in the Song Dynasty with the elegance of a modern city, together with Wanda Vista's brand feature of Personality, Finesse and bliss, the design fully elaborates the ubiquitous elegance of a millennial ancient capital. The design integrates ancient relique and modern urban style. Detail designs in antique flavour can be found all over the hotel so that the guests can feel the elegance and vogue of the 1000-year-old city.

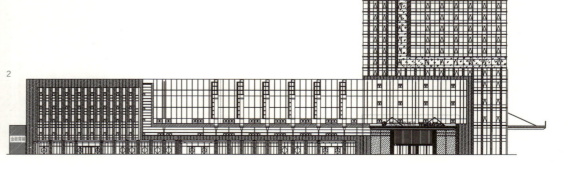

1 郑州万达文华酒店总平面图
2 郑州万达文华酒店立面图
3 郑州万达文华酒店外立面
4 郑州万达文华酒店夜景

酒店景观

根据建筑色彩,结合建筑立面特色的古典以及现代元素,打造具有鲜明艺术性和浓厚文化内涵的内外渗透的开放式景观,通过合理舒适的近人尺度,以及铺装、植物、水景、灯光等细节,营造精致华丽浪漫的酒店景观空间。

LANDSCAPE DESIGN

Based on the colors of the building, as well as featured classic and modern elements in the facade, we try to achieve open landscape with vivid artistry and strong cultural connotation. Delicate, splendid, romantic hotel landscape is achieved by means of reasonable, comfortable measures and details like paving, plants, waterscape, lighting, etc.

5 郑州万达文华酒店入口
6 郑州万达文华酒店外立面
7 郑州万达文华酒店水景
8 郑州万达文华酒店雨棚夜景

酒店夜景
NIGHTSCAPE DESIGN

03

WANDA VISTA URUMQI
乌鲁木齐万达文华酒店

OPENED ON: 12th AUGUST, 2016
LOCATION: URUMQI, XINJIANG UYGHUR AUTONOMOUS REGION
LAND AREA: 38,600 SQUARE METERS

开业时间：2016 / 08 / 12
开业地点：新疆维吾尔自治区 / 乌鲁木齐市
建筑面积：3.86万平方米

1 乌鲁木齐万达文华酒店总平面图
2 乌鲁木齐万达文华酒店立面图
3 乌鲁木齐万达文华酒店外立面
4 乌鲁木齐万达文华酒店外立面

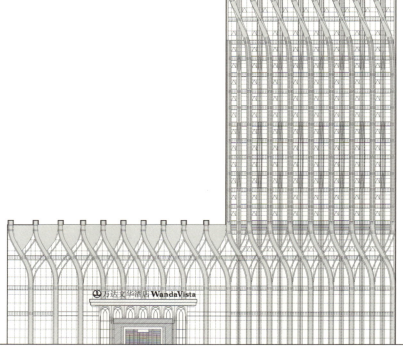

2

酒店概况

乌鲁木齐万达文华酒店坐落于乌鲁木齐市经济技术开发区核心地带，毗邻乌鲁木齐万达广场。酒店交通十分便捷，位于高铁新客站附近，距离乌鲁木齐地窝堡国际机场及乌鲁木齐火车南站分别15分钟车程。酒店拥有271间典雅舒适的客房与套房；设有中餐厅、全日餐厅、特色餐厅及大堂酒廊。1200平方米无柱大宴会厅，层高10米，配备90平方米的LED幕墙；并有7间不同规模的多功能厅和会议室。这些会议厅及会议室，结合会议活动设施，是举办各种宴会及活动的理想场所。

PROJECT OVERVIEW

The hotel is located in the center of Urumqi Economic and Technological Development Zone and next to Urumqi Wanda Plaza. Near the new high speed rail station, the hotel can be easily accessed. It takes 15 minutes' drive to respectively go to Diwobao International Airport and south Urumqi Railway Station. There are 271 elegant, comfortable guest rooms and suites. Chinese restaurant, all day dining, specialty food restaurant and bar are available. There is 1200 square meters pillarless banquet room with 10m storey height and a 90 square meters LED curtain wall equipped. There are 7 multi-purpose halls and meeting rooms in different sizes. Equipped with luxury meeting facilities, these meeting rooms are ideal for banquets and events.

1

3

酒店外装

乌鲁木齐万达文华酒店设计风格偏重现代与简练，设计元素从伊斯兰文化中提取精华，整体色调以高雅灰为主，并以蓝色为跳色贯穿其中。酒店设计主要元素采用新疆极具代表性的天山山脉及雪莲花形态，并将其运用于整个酒店的设计中，充分体现新疆本地的民族特色。

FACADE DESIGN

The style of the hotel is modern and simple. Islamic culture essence is used as the design elements. The building is in the main tone of elegant gray and is dotted with blue. The main elements in the design are the representative Tianshan Mountain and snow lotus flowers. They are used in the overall hotel design to fully represent local ethnic characteristics.

酒店景观

雄伟壮观、庄严神秘的天山是乌鲁木齐的守护神，方案以"天山"为设计雏形，以银辉的雪峰为主题，融入雪莲等元素深化人们心中的天山情结，以和建筑立面相匹配的设计元素来融入整体的场地设计。模拟天山自然地貌与山涧峡谷的空间感受，布置绿化、铺装、水景、小品等多重景观，将流动的弧线和转换的折线叠加形成空间的多样层次。

LANDSCAPE DESIGN

The grand, solemn and mysterious Tianshan Mountain is the guardian of Urumqi. Therefore, the design is based on "Tianshan Mountain". The shinning snow mountain is the theme, and elements like snow lotus flower are added to comfort the people's Tianshan complex. Elements matching the facade are used to echo with the overall site design. We tried to simulate the natural land form, streams, and gorges of Tianshan Mountain, arranged landscapes like green lands, paving, waterscape, etc., and put flowing curves and changing folding lines together to form cascaded spaces.

5 乌鲁木齐万达文华酒店入口
6 乌鲁木齐万达文华酒店绿化
7 乌鲁木齐万达文华酒店绿化
8 乌鲁木齐万达文华酒店景观

04

WANDA REALM SIPING
四平万达嘉华酒店

OPENED ON: 1st JULY, 2016
LOCATION: SIPING, JILIN PROVINCE
LAND AREA: 32,400 SQUARE METERS

开业时间：2016 / 07 / 01
开业地点：吉林省 / 四平市
建筑面积：3.24万平方米

1 四平万达嘉华酒店外立面
2 四平万达嘉华酒店总平面图
3 四平万达嘉华酒店外立面

酒店概况

四平万达嘉华酒店是五星级酒店，坐落于四平市核心商圈——四平万达广场，毗邻四平市主干道——紫气大道和开发区大路，距离高速公路口或高铁站仅15分钟车程。酒店设施齐全，拥有261间客房与套房以及能够同时容纳800人就餐的宴会厅与6间布局灵活的多功能厅，还配有嘉华行政酒廊、大堂酒廊、全日餐厅、日本餐厅、中餐厅、室内恒温游泳池及健身中心等设施。

PROJECT OVERVIEW

Wanda Realm Siping is a five-star hotel. It is located at Siping Wanda Plaza, the city's core business district, and next to Ziqi Road and Kaifaqu Road, the trunk roads of the city. It is only 15 minutes' drive from the hotel to the entrance of expressway or the high speed rail station. The hotel is well-equipped. There are 261 guest rooms and suites. The banquet hall can accommodate 800 guests for a dinner. There are 6 multi-purpose halls that can be flexibly arranged. Facilities like Realm executive lounge, lobby bar, all day dining, Japanese restaurant, Chinese restaurant, indoor swimming pool, gym are also available.

酒店外装

四平万达嘉华酒店采用竖向线条作为基本元素，立面装饰柱从飘扬的旗帜中提取，追求均衡的比例控制——"三段式"使得整个立面产生挺拔高耸的效果。酒店立面肌理具有较强的空间指向性，突出倾斜向上的态势，彰显建筑的张力和建筑的形体力量美感，并且准确地传达"英雄城"所独具的积极向上的城市动力！

FACADE DESIGN

Vertical lines are used as basic element. The decorative columns at the facade are derived from fluttering flag. Balanced control is pursued - The "tripartite composition" offers tall and straight facade. The hotel's facade texture offers relatively strong spacial orientation. It emphasizes an upward trend, shows the building's tension and beautiful shape so that it can accurately convey the "hero city" unique positive, upward power that is possessed only by the city.

4

4 四平万达嘉华酒店入口
5 四平万达嘉华酒店绿化
6 四平万达嘉华酒店夜景

5

酒店景观

四平市是满族的主要发祥地和聚居地，是满族的"祖宗肇兴之所"。在这块"贡山"、"贡河"相间环抱的沃土上，伊通满族在这里繁衍生息。作为满族传统文化缩影的"旗袍"不仅延续了当地历史文脉，也引领设计和时尚的潮流。景观设计以"旗袍"的图案和婀娜的曲线作为灵感的源泉。种植、铺装、小品等均采用柔滑的折线造型，既传统又现代，使人能够在商务休闲的空间中放松身心。

LANDSCAPE DESIGN

Siping is the cradle and habitat of Manchu, and the place where Manchu started their prosperity. On the fertile land encircled by "Gong Mountain" and "Gong River", Yitong Machu thrived As microcosm of Manchu's traditional culture, "Chipao" (Cheongsam) not only passes on local historical context, but also leads the trend of design and vogue. The patterns on "Chipao" and beautiful curves inspire the landscape design. Planting, paving, small landscape, etc. are modeled with smooth curves, which are both traditional and modern. Guests can be relaxed in the leisurely business space.

PART D WANDA HOTEL
万达酒店

酒店夜景
NIGHTSCAPE DESIGN

05

WANDA REALM XINING
西宁万达嘉华酒店

OPENED ON: 29th JULY, 2016
LOCATION: XINING, QINGHAI PROVINCE
LAND AREA: 41,500 SQUARE METERS

开业时间：2016 / 07 / 29
开业地点：青海省 / 西宁市
建筑面积：4.15万平方米

1 西宁万达嘉华酒店鸟瞰图
2 西宁万达嘉华酒店总平面图
3 西宁万达嘉华酒店外立面

酒店概况

西宁万达嘉华酒店为五星级酒店，坐落于西宁国家级新区海湖新区，毗邻西宁万达广场，距离西宁火车站30分钟车程，距西宁机场35分钟。酒店共拥有309间温馨典雅的客房与套房，及西宁最大的、100平方米超宽LED屏的1400平方米无柱大宴会厅；还配备有多种净高10米的中、小型会议室及6个多功能会议厅。餐饮方面有"品珍"中餐厅、特色餐厅、"美食汇"全日制西餐厅及大堂酒廊等。酒店拥有标准室内恒温游泳池、健身中心、桑拿房及美容美发等休闲娱乐功能。

PROJECT OVERVIEW

Wanda Realm Xining is a five-star hotel. It is located in Haihu New District, a new national district, and is next to Xining Wanda Plaza. It takes 30 minutes' drive to Xining Railway Station and 35 minutes' drive to Xining Airport. The hotel has 309 comfortable and elegant guest rooms and suites, and a 1,400 square meters pillarless banquet hall, the largest one in Xining, with 100 square meters wide LED screen. There are also several 10m high, middle- and small-sized meeting rooms and 6 multi-function conference halls. "Zhen" Chinese restaurant, specialty restaurant, All day dining western restaurant "Café Reign", lobby bar, etc. are available. There are standard indoor swimming pool, gym, sauna, beauty salon, and other facilities of leisure and recreation.

酒店外装

西宁万达嘉华酒店外立面设计灵感来源对地方文化的挖掘，取材于随风飘动的"藏族经幡"。酒店立面上流畅而简洁的曲线线条，就像飘荡在风中的祈愿幡，仿佛也承载着人们的希望和祝福！这一条条简洁的曲线，结合成组，形成韵律感和节奏感，似随风飘荡的五彩旗在空中舞动出美妙旋律。

FACADE DESIGN

The design of Wanda Realm Xining is inspired by local culture, and is based on "Tibetan prayer flags" fluttering in the wind. The smooth and simple curves on the facade are like the prayer flags fluttering in the wind, carrying on people's hope and blessing! The simple curves are grouped to give the sense of rhythm, just like wonderful melodies that those colorful flags are dancing with.

酒店景观

西宁是我国多民族聚居地之一，除汉族外，藏族、回族、土族以及撒拉族和蒙古族都世代生活于此，使得西宁彰显包容、勤奋、开放、创新的城市文化特点。西宁万达嘉华酒店以藏族文化中代表美好祝福的"哈达"作为景观元素，种植语言、景观雕塑、场地铺装，无一不以曲线律动的形式欢迎各民族同胞莅临。

LANDSCAPE DESIGN

Xining is one of the cities where several nationalities live. Besides the Hans, the Tibetans, Huis, Tues, Salars and Mongolians live here for generations so that Xining is outstanding for its inclusiveness, diligence, openness and creativity. "Hatha", which represents blessing in Tibetan culture, is a landscape element for the hotel. Without exception, the planting language, landscape sculpture and ground paving greet guests from all nationalities with curves.

6

4 西宁万达嘉华酒店外立面
5 西宁万达嘉华酒店入口
6 西宁万达嘉华酒店水景
7 西宁万达嘉华酒店夜景

酒店夜景
NIGHTSCAPE DESIGN

7

06

WANDA REALM BOZHOU
亳州万达嘉华酒店

OPENED ON : 12th AUGUST, 2016
LOCATION : BOZHOU, ANHUI PROVINCE
FLOOR AREA : 32,300 SQUARE METERS

开业时间 : 2016 / 08 / 12
开业地点 : 安徽省 / 亳州市
建筑面积 : 3.23万平方米

1 亳州万达嘉华酒店总平面图
2 亳州万达嘉华酒店外立面

酒店概况

亳州万达嘉华酒店是五星酒店，坐落于亳州市南部中央商务区，毗邻大型综合购物中心亳州万达广场；距离亳州火车站10分钟车程，距离邻近的阜阳机场1小时车程。酒店拥有263间宽敞舒适的客房与套房。其三楼有亳州市首屈一指的1100平方米无柱大宴会厅，配备了80平方米LED屏和一流的视听设备以及5个多功能会议厅。酒店餐饮方面配备有"品珍"中餐厅、"美食汇"全日制西餐厅、特色餐厅和大堂酒廊。酒店拥有标准室内恒温游泳池、健身中心、桑拿房及美容美发等休闲娱乐功能。

PROJECT OVERVIEW

Wanda Realm Bozhou is a five-star hotel. It is located in the BCD in the south of Bozhou, and is next to the large shopping center Bozhou Wanda Plaza. It is only 10 minutes' drive from Bozhou Railway Station, and 1 hour's drive from Fuyang Airport. The hotel has 263 spacious, comfortable guest rooms and suites. The 1100 square meters pillarless banquet hall on 3rd floor, which is the largest in Bozhou, is equipped with a 80 square meters LED screen, also the largest in the city, and top ranking audiovisual equipment. 5 multi-purpose meeting halls are also there on 2F. "Zhen" Chinese restaurant, all day dining western restaurant "Café Reign", featured restaurant, lobby bar, etc. are available in the hotel. There are standard indoor swimming pool, gym, sauna, beauty salon, and other facilities of leisure and recreation.

3 亳州万达嘉华酒店外立面
4 亳州万达嘉华酒店入口
5 亳州万达嘉华酒店景观
6 亳州万达嘉华酒店绿化

酒店外装

亳州万达嘉华酒店外立面设计灵感来源于本地地域文化特色和亳州作为"药都"的历史文化背景，借用采集药材的"背篓"形象加以艺术地再现，对"背篓编织"的构成元素进行重构——设计巧妙地提取"背篓"形态，使外立面竖向线条如同缠绕的藤条，与深灰色的玻璃产生了一种编织的肌理感觉——让酒店外立面在现代气息与地域文化中交织绽放！

FACADE DESIGN

Wanda Realm Bozhou's facade design is inspired by local culture and the historic background of Bozhou well known as the "city of medicines". "Pannier", which was used to collect herb medicines, is reproduced, and components for waving panniers are re-constructed - The design cleverly use the form of a "pannier" so that the façade's vertical lines are like winding canes, which produces the feeling of weaving together with dark grey glass - The facade of the hotel is so beautiful when it combines the modern sense and regional culture!

酒店景观

被誉为"天下道源,曹魏故里,中华药都"的亳州,具有悠久的文化传统。亳州万达嘉华酒店景观设计从亳州历史、人文、物产着手,穿插了古代走向未来的文化线索,绘成一幅具有亳州特色的涡河画卷。以"金樽迎宾"为主题的景观雕塑,坐落于酒店入口,传递了亳州人热情好客、淳朴的性格。

LANDSCAPE DESIGN

Known as "Cradle of Tao, the hometown of King Caocao, and the city of Chinese herb medicines", Bozhou has a long history and strong tradition. The landscape design of Wanda Realm Bozhou starts from the city's history, humanism and products, and inter-crosses the cultural clue from the past to the future, drawing a picture of the Wo River with local features. The landscape sculpture in theme of "greeting guests with gold bottles" is erected at the entrance of the hotel, showing Bozhou's hospitality.

07

WANDA REALM YIWU

义乌万达嘉华酒店

OPENED ON: 24th AUGUST, 2016
LOCATION: YIWU, ZHEJIANG PROVINCE
FLOOR AREA: 37,900 SQUARE METERS

开业时间： 2016 / 08 / 24
开业地点： 浙江义乌
建筑面积： 3.79万平方米

2

1 义乌万达嘉华酒店外立面
2 义乌万达嘉华酒店总平面图
3 义乌万达嘉华酒店景观外立面

酒店概况

义乌万达嘉华酒店是五星级酒店，坐落于时尚繁华的义乌万达广场，义乌江景由此一览无遗；酒店地理位置十分优越，距离高铁站14公里，距离义乌机场12公里。酒店拥有288间雅致的客房与套房，行政楼酒廊提供更高标准的豪华设施与个性服务。酒店大宴会厅层高8米，面积1200平方米，配备80平方米LED显示屏。大宴会厅配有先进的视听设备和多媒体设施，并可灵活分隔成三处独立空间分别使用。

PROJECT OVERVIEW

Wanda Realm Yiwu is a five-star hotel. It is at the fashionable, busy Yiwu Wanda Plaza. From the hotel, the view to the river scene is never obstructed. The hotel enjoys a good location only 14km away from high speed rail station, and 12km from Yiwu Airport. The hotel has 288 spacious, comfortable guest rooms and suites. The executive lounge is equipped with luxury facilities and offers customized services in higher standard. The hotel's grand banquet hall is 8m high and covers 1200 square meters. It is equipped with an 80 square meters LED screen. The grand banquet hall is equipped with advanced audiovisual equipment and multimedia facilities. It can be divided into three separate rooms.

酒店外装

义乌万达嘉华酒店建筑从树木形态中得到"枝条"的概念图案，以其成熟大气的建筑风格成为义乌开发区的新地标。与室内"水墨江南"主题意境相呼应展开设计，江南的烟雨空蒙、雅致、精细以及书香意气的情调被应用到了酒店从外到内的空间，体现出江南的灵秀细致、典雅自然。

FACADE DESIGN

The conceptional pattern of "tree branch" is conceived from the shape of a tree. It becomes the new landmark at Yiwu Development Zone for Its mature and magnificent architectural style. The design matches with the interior conceptual idea of "picturesque Jiangnan". Jiangnan's haziness, elegance and delicacy, as well as literary and poetic ambience, are applied all over the hotel, showing the delicate beauty, finesse and elegant naturalness of Jiangnan.

1

4

4 义乌万达嘉华酒店入口雨棚
5 义乌万达嘉华酒店景观
6 义乌万达嘉华酒店夜景
7 义乌万达嘉华酒店夜景

5

酒店景观

酒店建筑从树木形态中得到"枝条"概念图案取得灵感。景观风格承接建筑立面，以"溪流的肌理"在景观铺装和小品上加以诠释，使景观成为建筑的承载。义乌江及其支流构成了义乌的城市肌理，借用这一自然元素，整体沿用溪水流动的纹理作为铺装，以花坛和树木铺地模拟水中卵石，激荡出富有流线感的景观效果；提取"义乌道情戏"的伴奏乐器"鱼骨"和"百子花灯"为元素加以提炼，形成特色灯柱和小品。

LANDSCAPE DESIGN

The architecture is inspired by the conceptional pattern of "branch" on a tree. The landscape's style follows suit with the facade. The "textile of a stream" is used to guide the landscape paving and small landscape so that the landscape will carry along the building. The Yiwu River and its branches become the city's texture. The natural element is borrowed for the building, which generally uses the texture of a flowing stream as the paving, and simulates pebbles in the water with paved garden and trees so as to create the effect of streamline. The featured lamp posts and small landscapes are conceived from "Fish bone", the accompaniment in "Yiwu Daoqing Opera", and the "Hundred Children Lantern".

酒店夜景
NIGHTSCAPE DESIGN

08

WANDA REALM SHANGRAO
上饶万达嘉华酒店

OPENED ON : 25th NOVEMBER, 2016
LOCATION : SHANGRAO, JIANGXI PROVINCE
FLOOR AREA : 36,000 SQUARE METERS

开业时间：2016 / 11 / 25
开业地点：江西省 / 上饶市
建筑面积：3.6万平方米

1

1 上饶万达嘉华酒店总平面图
2 上饶万达嘉华酒店外立面
3 上饶万达嘉华酒店入口

酒店概况

上饶万达嘉华酒店地处广信大道繁华商业地段，距离信江双塔公园步行仅需5分钟，驱车20分钟即达上饶火车站，距离上饶机场仅8千米，距灵山20公里，距三清山70公里，距婺源138公里，旅游资源丰富，交通便利。酒店共拥有313间舒适的客房与套房，面积由40平方米的标准间~235平方米的总统套房不等。酒店拥有4家餐厅和酒吧、1000平方米无柱式大宴会厅以及5个多功能会议室，可以筹办大型会议、商务宴请和婚宴。

OVERVIEW

Wanda Realm Shangrao is on the Guangxin Road, in the middle of a bustling commercial area. It's only 5 minutes' walk from Xinjiang Shuangta Park, 20-minute drive from Shangrao Railway Station, 8km from Shangrao Airport, 20km from Lingshan Mountain, 70km from Sanqing Mountain, and 138km to Wuyuan. There are rich tourism resources and convenient transportation. The hotel has 313 comfortable guest rooms and suite, including 40 square meters standard rooms and 235 square meters presidential suites. The hotel has 4 restaurants and bars, one 1000 square meters pillarless banquet hall, and 5 multi-function meeting rooms so that it can be the best place to host large meetings, commercial banquets and wedding banquets.

酒店外装

上饶万达酒店外立面设计从"上饶印象"出发，自三清山独特的石林雾海获取灵感，着眼于酒店外立面材质肌理的生动表现，无论是竖向线条的纹理设计，还是对回纹图案的演变运用，都是意在设计独有而醒目的形象，从而使之进一步地融入上饶的地域文化中。

PROJECT OVERVIEW

The hotel's facade design is originated from "an impression of Shangrao", and is inspired by the unique "stone forest and fog sea" at Sanqing Mountain. The design evokes vivid expression of the façade's material and texture. Both the design of vertical lines' texture and the evolution and application of frets are intended to design unique and striking image so that they can be further integrated into local culture.

2

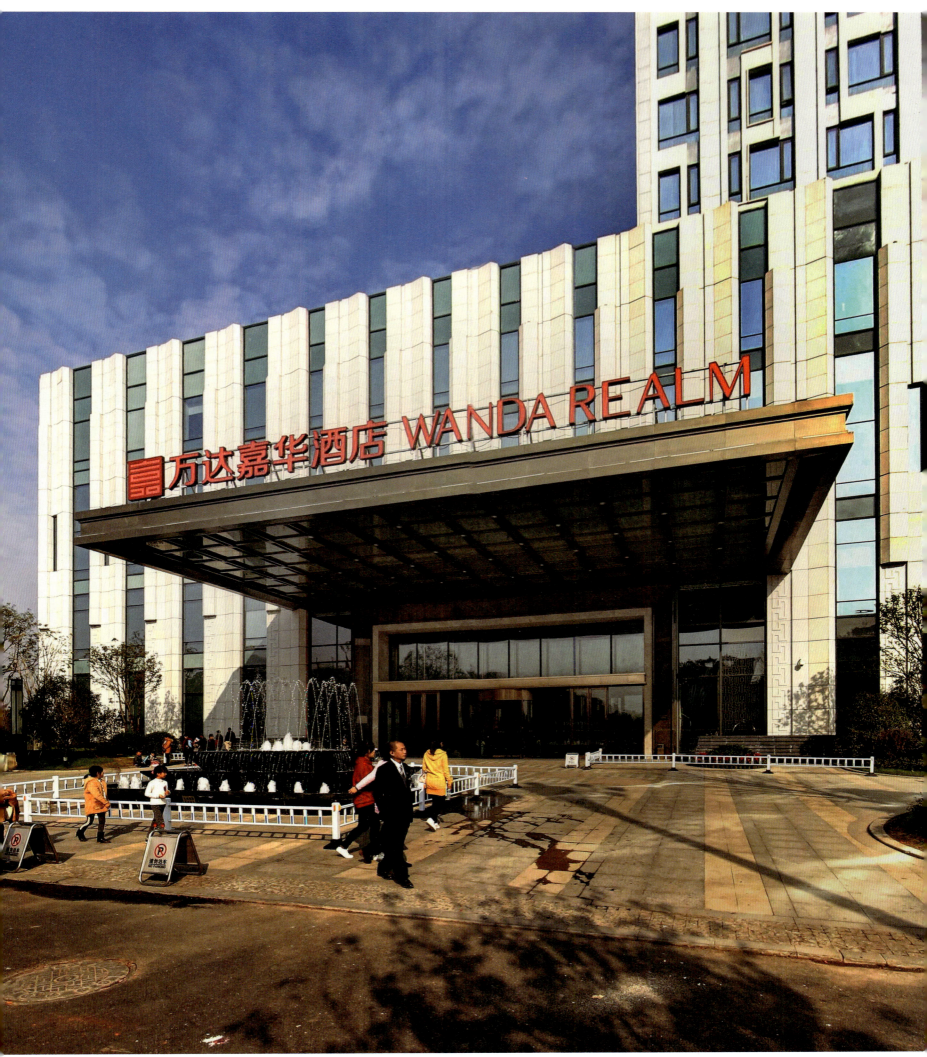

酒店景观

古人云"云以山为体,山以云为衣"。建筑构思取材于坐落于上饶的灵山,以山中之石抽象为建筑的立面元素,而景观理念则来源于灵山云海,好像"山中之云"一样环抱着场地和建筑。植物种植设计力求营造浓郁的商业气氛,通过色叶类植物以及开花缤纷的亚乔与灌木形成流动的动态肌理,为人们提供亲和、自然的活动环境。大商业外围主要以折线的铺装为主,增强人流的引导性,材料为花岗岩,色彩冷暖搭配。

LANDSCAPE DESIGN

Ancient people said that "Mountain is body of cloud while cloud is the clothes of a mountain". The architectural concept is originated from the Lingshan Mountain at Shangrao. The mountain stone is abstracted into the building's facade element. The landscape concept comes from the cloud sea of the mountain, so that it encircles the field and the buildings like "the cloud around a mountain". The plants at the commercial center are intended to create strong sense of commerce. The plants with colorful leaves and middle-sized blooming arbors and shrubs offer dynamic texture, and create comfortable, natural environment for activities. The large area surrounding the commerce is mainly paved like folded lines to offer better guidance to people flow. The materials are mainly granite with matching cold and warm colors.

4 上饶万达嘉华酒店水景
5 上饶万达嘉华酒店水景
6 上饶万达嘉华酒店喷泉
7 上饶万达嘉华酒店夜景
8 上饶万达嘉华酒店夜景

酒店夜景
NIGHTSCAPE DESIGN

销售类物业

PROPERTIES FOR SALE

01
EXHIBITION CENTER OF WANDA CITY CHONGQING
重庆万达城展示中心

OPENED ON : 23rd SEPTEMBER, 2016
LOCATION : SHAPINGBA DISTRICT, CHONGQING
LAND AREA : 2.1 HECTARES
FLOOR AREA : 5,700 SQUARE METERS

开业时间：2016 / 09 / 23
开业地点：重庆市 / 沙坪坝区
占地面积：2.1 公顷
建筑面积：5700 平方米

1 重庆万达城展示中心景观
2 重庆万达城展示中心总平面图
3 重庆万达城展示中心鸟瞰图
4 重庆万达城展示中心俯视图

展示中心概述

重庆万达展示中心——"山茶花"——精彩绽放重庆，既代表了重庆万达文化旅游城的整体形象、品质，又体现了重庆历史文化的特点及精髓。展示中心位于重庆市沙坪坝区西永园区，总用地面积2.1公顷，总建筑面积5700平方米，由万达文化旅游城超大型沙盘模型、5D体验厅及办公会议等功能组成。景观环境改造面积达10万平方米。

OVERVIEW OF EXHIBITION CENTER

The camellia-shaped Exhibition Center of Wanda City Chongqing has landed at Chongqing. The Center both represents the overall image and quality of Chongqing Wanda Cultural Tourism City and demonstrates the characteristics and essence of Chongqing's history and culture. It is located at Chongqing Xiyong Micro-Electronics Industrial Park, Shapingba District, with a gross land area of 2.1 hectares and a total floor area of 5,700 square meters. The Center consists of an extra large building model of Wanda Cultural Tourism City, a 5D experience hall, offices, meeting rooms and much more. The landscape renovation area reaches 100,000 square meters.

1

2

PART E PROPERTIES FOR SALE 销售类物业   215

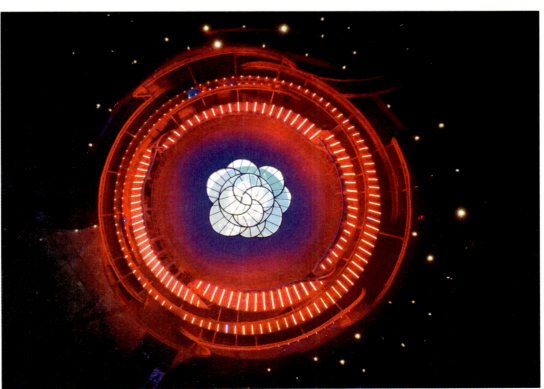

5 重庆万达城展示中心外立面
6 重庆万达城展示中心建筑结构图
7 重庆万达城展示中心外立面
8 重庆万达城展示中心外立面细节
9 重庆万达城展示中心夜景

展示中心建筑

设计理念——"山茶花"作为重庆市市花，代表勇敢拼搏的意义，表现重庆人民的精神内涵。设计巧用旋转叠加大小不一的花瓣，塑造了一朵行将绽放的"山茶花"；山城的绿地如花叶拥簇着花朵，开放于龙凤河边，花随风摇曳，陶醉于满园春色。

建筑空间——室内外空间变化丰富，移步异景。利用地形高差将二层直接与样板间相连，两层花瓣中加入观光外廊和休息平台。在建筑东北、西北两侧外廊均可以到达入口，直接进入展示中心室内二层。展示中心内部中央设有山茶花造型特色穹顶，与建筑主题相呼应。三层花瓣穹顶，在不同灯光效果的映衬下，呈现出茶花瓣的轻盈感和通透感。

建筑立面——设计提取山茶花整体的形状以及花瓣形态整合、抽象、旋转、叠加，使其富有动感。四层花瓣使用彩釉玻璃以及穿孔铝板，配合红色渐变褪晕，逐层变化形成丰富的立面形态。每层花瓣添加纵向纹理，增强立面细节效果；每逢夜晚，建筑通过色彩的渐变褪晕形成迷人的立面效果。

BUILDING OF EXHIBITION CENTER

Design Idea – Camellia, the city flower of Chongqing, implies courageous struggle, which is also the spiritual connotation the Chongqing people possess. The design skillfully creates a blooming camellia by rotating and overlaying petals of different sizes; the green space of Chongqing, seeming like leaves, gathers around the flower that is blossoming and swaying by the Longfeng River, rendering intoxicating spring colors.

Building Space - The interior and exterior spaces change frequently to such a way that "one step, one scenery" effect is delivered. By virtue of topographic difference, 2F is directly connected to the prototype room, with sightseeing verandah and rest platform being provided therein. In the northeast and northwest sides of the building, both verandahs there can lead to entrance and directly lead to 2F of the Center. In the middle of the interior, there is a featured camellia-shaped dome with three-layer petals, which echoes with the theme of the building and presents the light and crystalistic sense unique to Camellia pedals against colorful lights.

Building Facade – By way of integration, abstraction, rotation and superposition against the overall shape of the camellia and its petal form, the design endows the flower with dynamism. The four - layer petals made of colored glazed glasses and perforated aluminum plates boast of layered variation combined with red color gradient and produce abundant facade forms. Each layer also has vertical texture to highlight facade details; once night falls, the building will present a charming facade accompanying the color change.

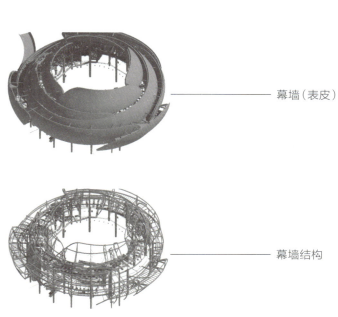
幕墙（表皮）

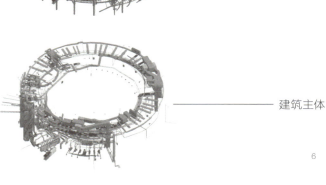
幕墙结构

建筑主体

  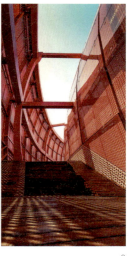

10

展示中心景观

展示中心被无边无际的绿色海洋包围，形成"万绿丛中一点红"的诗意画面。14万平方米的展示范围、2万平方米的河道、1400米长的景观大道、大量的乡土树种，展现了草木茂盛、山花烂漫的活力景观。设计在充分研究地势地貌的基础上，将大山大水自然景观与展示中心、示范体验融为一体。

LANDSCAPE OF EXHIBITION CENTER

Embracing by a vast green space sea, the Center displays a poetic picture of a single red flower in the midst of thick foliage. The 140,000 square meters exhibition area, 20,000 square meters river channel, 1,400m landscape avenue and plentiful local trees together contribute to a dynamic landscape full of lush trees and beautiful flowers. Through a complete research on the terrain and landform, the design fully presents an integration of natural landscape of water and mountains into the exhibition center and demonstration area.

10 重庆万达城展示中心水景
11 重庆万达城展示中心景观规划图
12 重庆万达城展示中心景观栈道
13 重庆万达城展示中心绿化

14

17

15

16

14 重庆万达城展示中心接待台
15 重庆万达城展示中心洽谈区
16 重庆万达城展示中心洽谈区
17 重庆万达城展示中心洽谈区
18 重庆万达城展示中心大厅

展示中心内装

展示中心内装运用色彩的叠加、晕染，塑造了涌动的层次关系，具有步移景易的效果。室内室外呼应，在精心设置的照明映衬下，3层穹顶拉膜花瓣通透而优雅地绽放。

INTERIOR OF EXHIBITION CENTER

The interior design builds a flowing level relationship with color superposition and shading, creating the effect of one step, one scenery. Echoing with the facade design, the interior three-layer dome's film-drawing petals bloom gracefully and crystally under the well-arranged lighting.

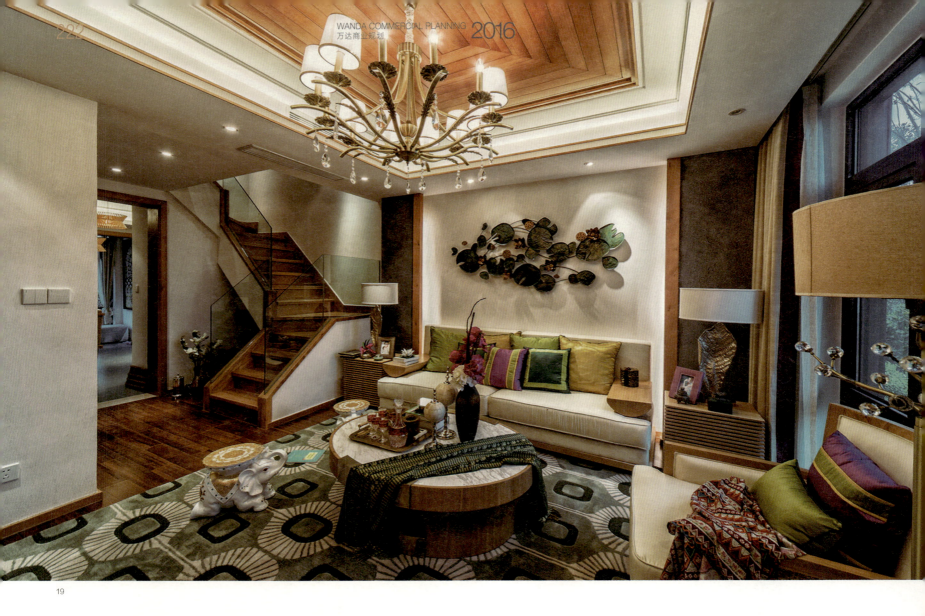

别墅样板间

东南亚风格别墅样板间,以亮白基调的墙纸、文化肌理感的石材,结合实木线条贯穿整个空间。在软装搭配上运用了棉麻、藤编家具材质,点缀艳丽的东南亚风情装饰品,营造出自然、异域风情的空间氛围。

VILLA PROTOTYPE ROOM

The villa prototype room adopts Southeast Asian style and it fills the space with bright white wall paper, stone with cultural texture and solid wood moulding. In addition to the showy Southeast Asian adornments, the soft decoration also picks up some cotton & linen and rattan furniture, presenting a natural and exotic atmosphere.

19 重庆万达城展示中心别墅样板间客厅
20 重庆万达城展示中心别墅样板间餐厅
21 重庆万达城展示中心别墅样板间卧室
22 重庆万达城展示中心高层样板间客厅
23 重庆万达城展示中心高层样板间餐厅、客厅
24 重庆万达城展示中心高层样板间书房

高层样板间

高层新中式样板间以中国水墨画为主题元素，结合现代工艺技术及设施（包括地面的地毯、墙面的装饰画、客厅茶几及装饰品）表达传统东方美感；再搭配现代装饰、家具，营造出古今交融的空间氛围。

HIGH-RISE PROTOTYPE ROOM

The high-rise prototype room observes new Chinese style and it largely applies the element of Chinese ink painting. The design interprets the traditional Oriental beauty by adopting modern technologies and facilities (including the carpet on floor, the decorative painting on wall, the tea table in living room and ornaments); it also builds up a combined ancient and modern style space with provision of modern adornment and furniture.

|02

EXHIBITION CENTER OF WANDA NO.1 CHENGDU

成都万达一号展示中心

OPENED ON : 18th AUGUST, 2016
LOCATION : CHENGDU, SICHUAN PROVINCE
LAND AREA : 2,007 SQUARE METERS
FLOOR AREA : 3,140 SQUARE METERS

开业时间: 2016 / 08 / 18
开业地点: 四川省 / 成都市
占地面积: 2007 平方米
建筑面积: 3140 平方米

1 成都万达一号展示中心外立面
2 成都万达一号展示中心总平面图
3 成都万达一号展示中心立面图

展示中心概述

成都万达一号,坐落于成都市天府新区武汉东路。展示中心提取山、水及四川传统民居的元素,将人文、自然引入其中,使建筑体与自然交织——人在其中,宛如画中——营造"由景入境"的古韵主题。通过"抽象"和"写意"的主题呈现,将"醉美"的山水景色进行"去图案化",采用抽象的"线构"方式来进行再现——巧借人工,抽象自然——既表现了展示中心的现代性,又将成都特有的山水景色融于其中。

OVERVIEW OF EXHIBITION CENTER

Wanda No.1 Chengdu is located on East Wuhan Road, Tianfu New District, Chengdu. The exhibition center adds culture and nature to the elements of mountains, waters and traditional folk houses in Sichuan Province, enabling perfect harmony between building and nature and highlighting the archaic theme of evoking artistic conception by scenery so that visitors are wandering in picturesque scene. The design employs abstract and spontaneous presentation to enable de-patterning of the intoxicating landscape; moreover, it adopts linear structure for a kind of representation-artificially abstracting nature. As a result, the Center incorporates not just modern features, but also unique landscape in Chengdu.

展示中心建筑

成都万达一号展示中心的立面构思基于成都休闲、时尚之都的文化基调，虚实对比恰当；色彩搭配协调；线条流畅大方，与整体建筑风格相互映衬。

BUILDING OF EXHIBITION CENTER

The facade design of the Center is based on the cultural keynote of Chengdu - a leisure and fashion city. It boasts of sound virtual-real comparison, coordinated color collocation, smooth and generous mouldings, and a style in harmony with buildings.

4 成都万达一号展示中心夜景
5 成都万达一号展示中心外立面
6 成都万达一号展示中心外立面
7 成都万达一号展示中心夜景

8

展示中心景观

成都万达一号景观设计的目的：打造天府新区"山水怡和，低调奢华，生态颐养，文墨情怀"高级住宅形象，在空间、文化、细节方面深入刻画，形成如下特点：隐（绿色围合），院（内向庭院），水（水面阔达），廊（巧奥空间），精（精雕细琢），雅（简朴优雅），巧（借景抒情），成为展示中国文化与礼仪新标杆、品质栖居新典范、健康生态新样板、文雅情怀新形象。

LANDSCAPE OF EXHIBITION CENTER

The landscape design of Wanda No.1 Chengdu aims to build the high-grade residence image in Tianfu New District, which is characterized by landscape harmony, modest luxury, ecological comfort and cultural sensation. It thus delves into space, culture and details to contain the following characteristics: seclusion (green enclosure), courtyard (inward courtyard), water (vast water), corridor (ingenious space), refinement (well carving), elegance (simple & elegant); skillfulness (scenery-based emotion). The landscape will emerge as a new benchmark in displaying Chinese culture and etiquette, a new model of quality residence, a new sample of healthy ecology and a new image of elegance.

9

8 成都万达一号展示中心入口广场
9 成都万达一号展示中心夜景
10 成都万达一号展示中心景观亭
11 成都万达一号展示中心水景

12

13

12 成都万达一号展示中心接待台
13 成都万达一号展示中心屏风
14 成都万达一号展示中心大厅
15 成都万达一号展示中心沙盘
16 成都万达一号展示中心洽谈区

14

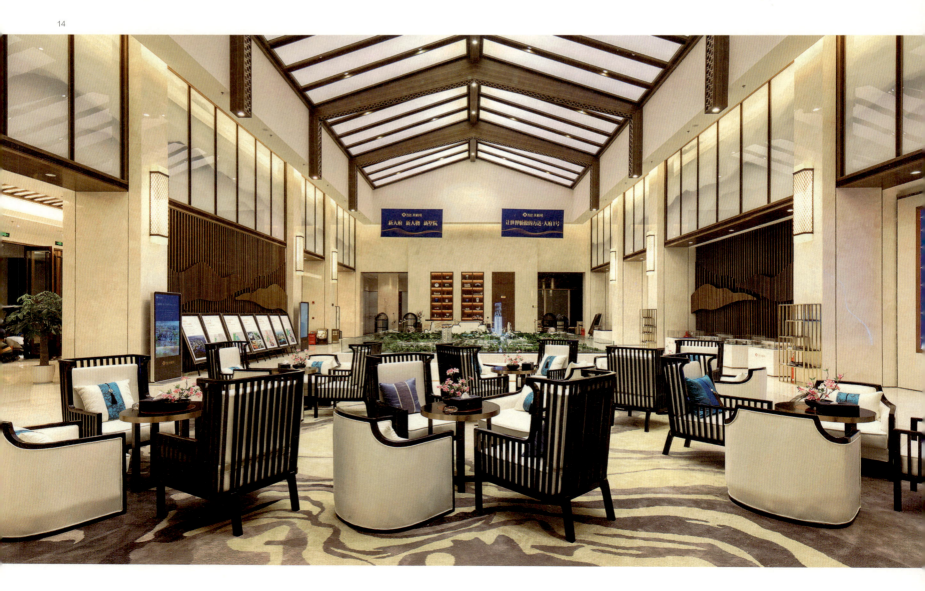

展示中心内装

成都万达一号售楼处的内装设计有别于万达以往售楼处的设计，采用以意境营造为主，将功能置于其中的处理。售楼处空间造型结合地域特色，以中国传统的生态园林建筑为基础；室内装饰元素以四川秀美山水的形态为依托，将自然韵味引入室内空间，使空间与自然交织，营造出禅意、雅致、悠闲的空间氛围，强调文化感受，打造出一处"不以形奢，宛在画中"的体验式商业空间。

INTERIOR OF EXHIBITION CENTER

The sales office interior design of Wanda No.1 Chengdu stands out from the remaining Wanda's sales offices, as it includes functions into artistic conception. The space shape absorbs regional features and mainly relies on traditional Chinese ecological garden architecture. The interior decoration draws elements from beautiful scenery in Sichuan and introduces them into the interior space. When space and nature are put together, a graceful and leisurely atmosphere full of Zen ideology is felt, and an experiential commercial space that defies luxury while highlights cultural perception and picturesque environment is taking shape.

03

HEPING EXHIBITION CENTER OF WANDA CITY GUILIN
桂林万达城和平展示中心

OPENED ON: 24th SEPTEMBER, 2016
LOCATION: QIXING DISTRICT, GUILIN
LAND AREA: 4.5 HECTARES
FLOOR AREA: 3,600 SQUARE METERS

开业时间： 2016 / 09 / 24
开业地点： 桂林市 / 七星区
占地面积： 4.5 公顷
建筑面积： 3600 平方米

1 桂林万达城和平展示中心外立面
2 桂林万达城和平展示中心总平面图
3 桂林万达城和平展示中心立面图
4 桂林万达城和平展示中心鸟瞰图

1

展示中心概述

桂林万达城和平展示中心以"桂林山水"为设计主题，既代表了桂林万达文化旅游城的整体形象、品质，也体现了桂林"山水文化"的特点及精髓。展示中心立于桂林市七星区，总用地面积4.5公顷，总建筑面积3600平方米，由万达文化旅游城超大型沙盘模型、接待及办公会议等功能组成。

OVERVIEW OF EXHIBITION CENTER

Heping Exhibition Center of Wanda City Guilin is designed centering on Guilin Landscape. The Center both represents the overall image and quality of Guilin Wanda Cultural Tourism City and demonstrates the characteristics and essence of Guilin's landscape culture. It is ocated at Qixing District, with a gross land area of 4.5 hectares and a total floor area of 3,600 square meters. The Center consists of an extra large building model of Wanda Cultural Tourism City, a reception, offices, meeting rooms and much more.

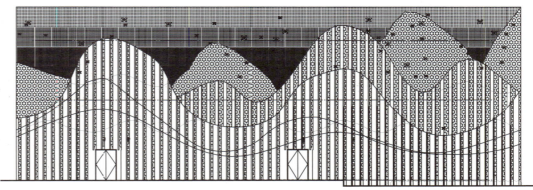

5 桂林万达城和平展示中心外立面
6 桂林万达城和平展示中心外立面
7 桂林万达城和平展示中心外立面

展示中心建筑

设计理念——"桂林山水"、"甲天下"。"桂林山水'作为中国山水的代表、世界自然遗产,是桂林文化的象征。设计塑造了一副缓缓展开的山水画卷——从山水到建筑,含蓄地表达出自然的景色:复杂而又节制、自由而有韵律。

建筑立面——"桂林山水甲天下,玉碧罗青意可参"。纯净的玻璃盒子,透过玻璃幕墙的处理(同色系的玻璃片拼贴)抽象化表达对山水的理解;夜晚"四季景色"变化、"九马画山"又使山水灵动起来,亦静亦动……

BUILDING OF EXHIBITION CENTER

Design Idea – The mountains and waters of Guilin are the finest under heaven. Guilin Landscape, as the representative of China's landscape and the World Natural Heritage, symbolizes the culture of Guilin. The design moulds a slowly unfolding landscape scroll, implying the natural scenery from landscape to architecture-complex and temperate, free and rhythmic.

Building Facade – The mountains and waters of Guilin are the finest under heaven and are worth appreciating. The facade utilizes the crystal glass boxes to abstract the understanding of landscape through special design of glass curtain wall (piecing together glasses of the same color scheme); the scenery of four seasons at night and the nine horses contribute to a kinetic landscape, static and dynamic ..

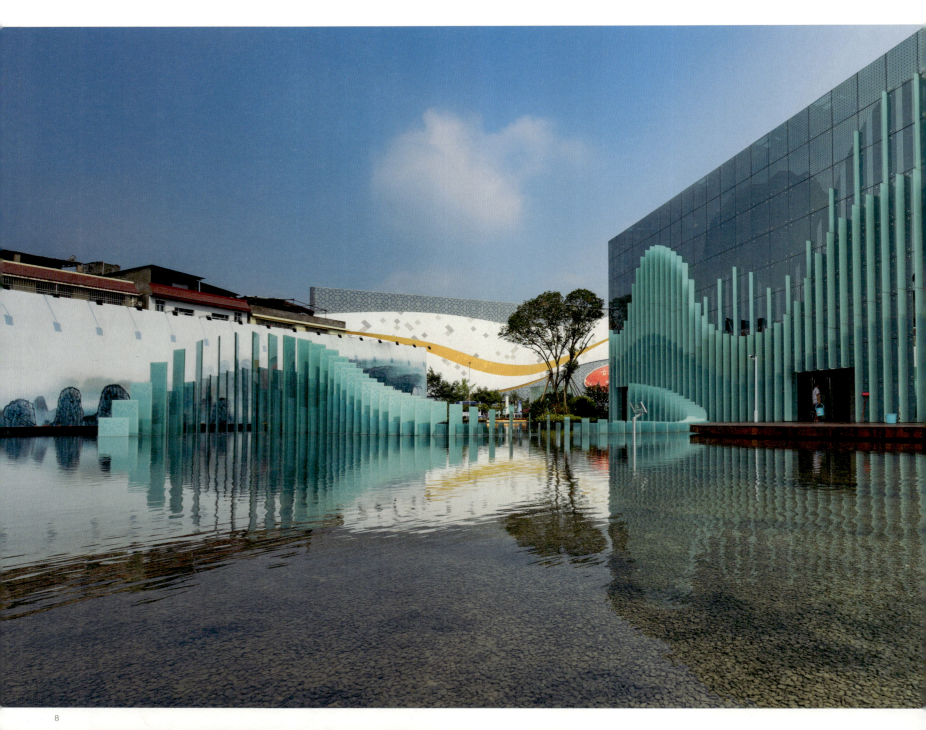

展示中心景观

展示中心被一池静水包围，衬托于绿水之上。水池中延续建筑元素的玻璃片山水造型，使山水意向舒展开来；曲线的叠水、水中的休息椅子、竹材料艺术运用、木平台等让参观者一步一景，无处不在"赏山、游水"动态变化，浓缩桂林山水于画中。

LANDSCAPE OF EXHIBITION CENTER

The Center is embraced and supported by a pool of static water. The pool still employs the glass-made landscape shape applied in the building to leisurely extend the scenery. The curved cascading, resting chair in the water and wooden platform render visitors a changing and ubiquitous scenery experience.

8 桂林万达城和平展示中心水景
9 桂林万达城和平展示中心水景
10 桂林万达城和平展示中心竹连廊
11 桂林万达城和平展示中心竹连廊
12 桂林万达城和平展示中心竹连廊夜景

13

14

13 桂林万达城和平展示中心大厅
14 桂林万达城和平展示中心贵宾室
15 桂林万达城和平展示中心书吧
16 桂林万达城和平展示中心洽谈区
17 桂林万达城和平展示中心VR体验区

展示中心内装

通过建筑"洞口"进入室内空间，感受"我们乘着木船，荡漾在漓江上，来观赏桂林的山水"。空间流动的线条赋予空间以灵魂，或静止、或流动、或理性、或感性，体块之间高低起伏的变化及迂回体现穿行的乐趣，在其中感受充满生机与感动的生活意象。

INTERIOR OF EXHIBITION CENTER

The nterior space is accessible via the building hole, feeling as if we are apprec ating Guilin mountains and waters along the Lijiang River in a wooden boat. The flowing lines inside enliven the space, making it static, moving, rational or emotional. The undulating and twisting blocks add the fun of passing through therein, experiencing a kind of vigorous and touched life image.

18

18 桂林万达城和平展示中心禅意风格样板间客厅
19 桂林万达城和平展示中心禅意风格样板间卧室
20 桂林万达城和平展示中心东南亚风格样板间户型图
21 桂林万达城和平展示中心东南亚风格样板间卧室
22 桂林万达城和平展示中心东南亚风格样板间卫生间
23 桂林万达城和平展示中心东南亚风格样板间客厅

禅意风格样板间

室内陈设色调延续木本色的基础色调，将肌理涂料、天然大理石等自然材质巧妙运用其中，让空间不仅有自然的拙朴感，还有禅意的清雅味道。软装融入清丽的翠绿色，与室外美景色隔空对望，悦见悠然禅境。

ZEN STYLE PROTOTYPE ROOM

Indoor display stresses on the timber color and skillfully adds such natural materials as texture paint and natural marble therein, rendering a naturally unadorned and Zen-style fresh space. While soft decoration integrates the bright green color, which echoes with the outdoor beauty afar and further hilights the carefree Zen state.

19

东南亚风格样板间

在造型上，该样板间以对称的木结构为主，营造出浓郁的热带风情。在色彩上，以温馨淡雅的中性色彩为主，局部点缀艳丽的绿色、紫色，自然温馨中不失热情华丽。在材质上，运用实木、硅藻泥等，演绎原始自然的热带风情。软装采用纯天然藤条、竹子、棉麻等材料来进行装饰，突出热带雨林的纯朴自然之美和浓郁的民族特色。

SOUTHEAST ASIA STYLE PROTOTYPE ROOM

In terms of shape, symmetrical wood structure is largely applied to build up a strong tropical flavor. In terms of color, warm and elegant neuter colors are dotted with showy green and purple colors, natural & sweet yet fiery & gorgeous. In terms of material, solid wood and diatom ooze are used to display primitively tropical atmosphere. In terms of soft decoration, pure natural cane, bamboo, cotton & linen are helpful to highlight the natural beauty of tropical rainforest and the strong national characteristics.

04

DEMONSTRATION AREA OF YONGCHUAN WANDA PALACE
永川万达华府示范区

OPENED ON : 7th APRIL, 2016
LOCATION : YONGCHUAN DISTRICT, CHONGQING
LAND AREA : 3,200 SQUARE METERS
FLOOR AREA : 2,200 SQUARE METERS

开业时间：2016 / 04 / 07
开业地点：重庆市 / 永川区
占地面积：3200 平方米
建筑面积：2200 平方米

1

示范区概况

永川万达华府作为永川万达广场的配套销售物业，建筑面积近46万平方米。住宅户型设计强调实用高效，主客活动空间相对独立分区，互不干扰。起居厅和餐厅分开设置，内外、洁污分离；窗户与阳台的设置充分考虑景观视野，让内外空间相互渗透。户型设计主要为一梯两户洋房构成，面积介于95~130平方米。

OVERVIEW OF DEMONSTRATION AREA

Yongchuan Wanda Palace is the supporting properties for sale of Yongchuan Wanda Plaza, and it covers a floor area of nearly 460,000 square meters. The house type adopts practical and efficient design. Host and guest activity spaces are distributed in independent and non-interfering zones; living room and dining room, interior and exterior, cleanliness and dirt are separated; the setting of window and balcony gives full consideration to landscape view, enabling a mix of internal and external spaces. The house type mainly includes bungalows with one stair and two households, with room area ranging between 95 square meters to 130 square meters.

PART E PROPERTIES FOR SALE
销售类物业

243

1 永川万达华府示范区总平面图
2 永川万达华府示范区建筑外立面
3 永川万达华府示范区建筑立面图

示范区建筑

洋房立面采用地中海风格，以全新的思维及手法追求简洁大气的美感，使该小区建筑从周围环境中脱颖而出。

BUILDING OF DEMONSTRATION AREA

The bungalow facade features Mediterranean style. It pursues a kind of simple & atmospheric aesthetics with brand-new thoughts and techniques, making the residential building stand out from the surrounding environment.

4 永川万达华府示范区建筑外立面
5 永川万达华府示范区建筑外立面
6 永川万达华府示范区建筑外立面

7

8

9

示范区景观

洋房示范区用地面积3200平方米,包含洋房样板间、样板小院以及公共环境景观区。洋房采用了地中海风格整体景观体系,呼应重庆人生活的悠闲、自由。设计采用鲜艳的色调、多变的材料、灵活的空间,使整相对狭小空间,呈现出了更加丰满的层次关系。对景借鉴地中海风格经典的景墙处理方式。在入口景墙的两侧,划分"去"与"回"两条路线,布置了四种风格样板私家庭院,通过景观节奏的变化营造印象深刻的参观体验。

7 永川万达华府示范区聚餐区
8 永川万达华府示范区庭院
9 永川万达华府示范区石板路
10 永川万达华府示范区庭院

LANDSCAPE OF DEMONSTRATION AREA

The villa demonstration area covers an area of 3,200 square meters, including villa prototype room, courtyard prototype and public environment landscape area. The villa largely applies a landscape system of the Mediterranean style, which echoes with the leisurely and carefree life in Chongqing. By virtue of bright colors, changeable materials and flexible spaces, the design endows the relatively small space with rich gradation. Inspired by the classic landscape wall design unique to the Mediterranean style, the opposite scenery design plans two paths (back and forth) on both sides of the entrance landscape wall. In this way, four private courtyards of different styles are provided, which bring visitors an impressive experience through the change of landscape rhythm.

洋房样板间

洋房样板间采用现代港式风格。设计色彩冷静、线条简单、空间秩序分明。顶棚与立面运用大量的金属明线收口。客厅、卧室的背景墙采用与墙面色彩近似,纯几何网格分线的硬包。墙面、地面、顶棚以及灯具器皿、家具陈设均以简洁的造型、纯洁的质地、精细的工艺为特征。少量的紫色、湛蓝色的软装布艺的介入,让整个暖色基调的空间变得清新透气。

BUNGALOW PROTOTYPE ROOM

Villa prototype room pursues modern Hong Kong style, featuring cool colors, simple mouldings and orderly space. Ceiling and facade largely apply metal surface lines for shell nosing; living room and bedroom use the background wall of the similar color with wall and hard decoration made of pure geometry grid; wall, floor, ceiling as well as lamps and furniture are all characterized by simple shape, pure texture, fine craft. Plus a handful of purple and sky blue fabrics, the warm color-based space seems to be fresh and crystal.

14

11 永川万达华府示范区洋房样板间入户大堂
12 永川万达华府示范区洋房样板间户型图
13 永川万达华府示范区洋房样板间书房
14 永川万达华府示范区洋房样板间卧室
15 永川万达华府示范区洋房样板间客厅

15

05

PROTOTYPE SHOP DEMONSTRATION AREA OF WANDA CITY CHENGDU
- GREEN TOWN & WATER STREET

成都万达城商铺样板示范区
——青城水街

OPENED ON : 12th JULY, 2016
LOCATION : DUJIANGYAN, CHONGQING
LAND AREA : 1.5 HECTARES
FLOOR AREA : 6,000 SQUARE METERS

开业时间： 2016 / 07 / 12
开业地点： 成都 / 都江堰市
占地面积： 1.5 公顷
建筑面积： 6000 平方米

2

1 成都万达城商铺样板示范区——青城水街商铺外立面
2 成都万达城商铺样板示范区——青城水街总平面图
3 成都万达城商铺样板示范区——青城水街水景
4 成都万达城商铺样板示范区——青城水街商铺门头

1

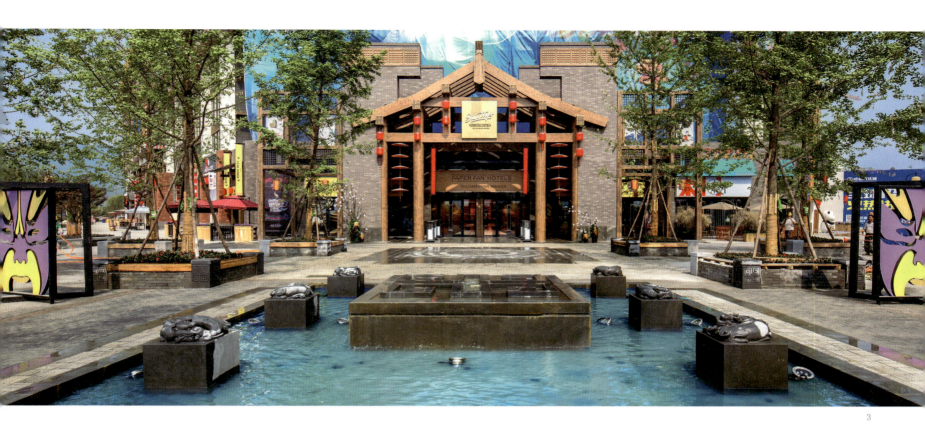

项目概况

成都万达城商铺样板示范区——青城水街项目位于都江堰市，为成都万达城一期，总用地面积1.5公顷，总建筑面积6000平方米，全部为一层临街商铺。示范区商铺层高6米，开间3.6～4.2米，进深15米左右，立面采用现代川西风格。

建筑规划

青城水街在规划上与用地条件紧密结合，沿道路走势延展布置，形成自然柔和的街道空间形态。建筑单体设计上，一方面充分考虑到商铺经营的需求，采用小面积商铺、灵活组合的方式，以满足不同商户的需求，另一方面，与住宅建筑相互融合，打造统一的整体街区形象。

PROJECT OVERVIEW

Prototype Shop Demonstration Area of Wanda City Chengdu - Green Town & Water Street belongs to Wanda City Chengdu, Phase I, in Dujiangyan, Chengdu, with a gross land area of 1.5 hectares and a gross floor area of 6,000 square meters. The demonstration area includes one-floor shops along street only, each being 6m high, 3.6m - 4.2m wide, and 15m deep with a modern Western Sichuan-style facade.

BUILDING PLANNING

The Project is closely combined with the land conditions in its planning, as it extends along street trend and presents a natural and gentle street space form. In terms of individual building, the design, on the one hand, issues small shops which enable flexible combination to meet the needs of different merchants, by taking full account of the operation demand of shops; on the other hand, the design is integrated with residential buildings to create a unified block image.

建筑立面

项目立面采用传统与现代相结合的方式，运用现代语言来表达传统的建筑风格，提炼出川西民居特有的元素，将其进行再创作。在非原汁原味的民居里，却能品味到浓浓的川西风味。

BUILDING FACADE

The Project facade represents traditional architectural style with modern language by way of a mixed traditional and modern approach. It re-creates the featured elements of Western Sichuan folk houses and brings them back to the building. As a result, one may discern the strong Western Sichuan flavor from these non-original folk houses.

5 成都万达城商铺样板示范区——青城水街商铺外立面
6 成都万达城商铺样板示范区——青城水街商铺外立面
7 成都万达城商铺样板示范区——青城水街商铺外立面
8 成都万达城商铺样板示范区——青城水街商铺门头

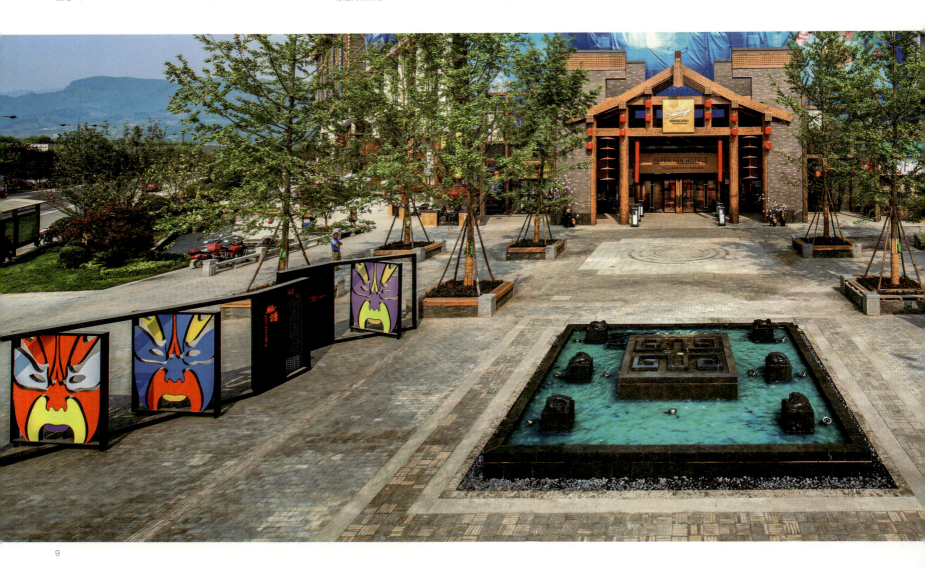

9

9 成都万达城商铺样板示范区——青城水街景观
10 成都万达城商铺样板示范区——青城水街景观小品
11 成都万达城商铺样板示范区——青城水街景观小品
12 成都万达城商铺样板示范区——青城水街景观小品
13 成都万达城商铺样板示范区——青城水街水景
14 成都万达城商铺样板示范区——青城水街景观小品

10

11

景观设计

景观设计以"青山堰水，山水画卷"为设计理念，商街形成水文化主题街区，以水文化为脉络，利用现有河道水源，渠道引流，满足下游灌溉功能的同时，形成特有的水街景观。水文化以时间为脉络，挖掘治水文化相关民间故事，表达人与水的从观到用的互动关系，形成川西传统与现代结合的风情水街。

LANDSCAPE DESIGN

The landscape design presents the design philosophy of "A Chinese Landscape Painting with Blue Mountains and Streams Around". The commercial street forms a theme block of water culture by using the existing river courses to irrigate the downstream and also to form waterscape street with vivid characteristics. The water culture applies the time as a sequence to explore the folk stories of water conservancy culture, indicate the interaction of human and water and build a style street in the combination of western Sichuan tradition and the modern time.

06

COMMERCE & RESIDENCE BLOCK OF ZHANGZHOU TAISHANG WANDA PLAZA

漳州台商万达广场商墅

OPENED ON: 30th NOVEMBER, 2016
LOCATION: ZHANGZHOU, FUJIAN PROVINCE
LAND AREA: 7,915 SQUARE METERS
FLOOR AREA: 17,500 SQUARE METERS

开业时间：2016 / 11 / 30
开业地点：福建省 / 漳州市
占地面积：7915 平方米
建筑面积：1.75 万平方米

1 漳州台商万达广场商墅总平面图
2 漳州台商万达广场商墅建筑立面图
3 漳州台商万达广场商墅大门

项目概况

漳州台商万达广场位于漳州市台商区核心区，交通便捷，通达性好，具有地标价值和土地价值，是地标性、综合性的"万达城市综合体"。本次改造将原住宅三期、四期底商十字街区，改造为拥有一个中心广场、两条百花古巷、四座墅府大门、别具闽南院落式风格的"商墅"街区。

PROJECT OVERVIEW

Ideally located in the core area of Taishang District, Zhangzhou, the Zhangzhou Taishang Wanda Plaza enjoys convenient transportation and easy access, and has both landmark value and land value. It is thus an iconic and integrated Wanda City Complex. The renovation intends to change the former cross-shaped block of Phase III and Phase IV commerce at the bottom into a commerce & residence block with Southern Fujian courtyard style, including one central plaza, two ancient alleys and four gates.

建筑规划

打造"入户院落+墅内中庭+屋顶花园+中央景观"的商墅产品，具有多附加值、多景观视角，成为万达首创的"别院商墅"，具有广泛的适应性。商墅独立成区，与已建成的住宅大区和谐共处，"一个中心广场、两条百花古巷、四座墅府大门"是规划的关键，迎来"里弄式"传统生活的气息。

ARCHITECTURAL PLANNING

The architecture plans to produce a "commerce & residence block" product featuring household "courtyard + villa atrium + roof garden + central landscape", with more added values and landscape views. It also strives to emerge as the first "courtyard-style commercial villa" initiated by Wanda, having extensive adaptability. The Block is independently constructed and in perfect harmony with the completed residential buildings. With stress put on "one central plaza, two ancient alleys and four gates", the planning tries to bring in a Alley-style traditional life atmosphere.

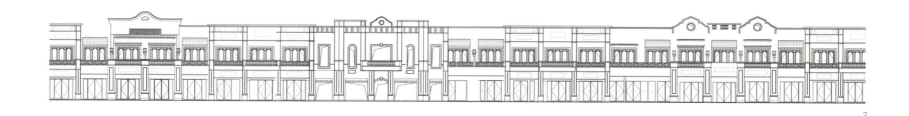

立面设计

商墅对外立面进行品质提升,新开间可与原立面开间一一对应,并通过任意拼接形成新的街区立面。通过建筑立面造型高低错落,结合闽南风韵的红砖、当地特色的石材和浮雕、闽南符号的马头墙等建筑构件,充分展示闽南古厝的民俗民风,打造闽南地域特色的品质街区。

FACADE DESIGN

The Block's facade has an improved quality. Though the new bay is exactly matched with the original facade bay, a new block facade is made available through random piecing. The design applies staggered facade shapes, red bricks unique to Southern Fujian, locally featured stones and reliefs, Ma Tau Wall (common in Southern Fujian) and other building components. These efforts fully demonstrate the folk customs of old-age house in Southern Fujian and innovatively build up a quality block with regional characteristics.

4 漳州台商万达广场商墅建筑外立面
5 漳州台商万达广场商墅建筑外立面
6 漳州台商万达广场商墅公共空间景观
7 漳州台商万达广场商墅公共空间景观

景观设计

以"闽南九里、百花古巷"为景观设计主题,营造温馨的邻里商墅空间氛围,串联起入口广场、社区大门、百花古巷、公共邻里中心、私家别院等行进序列;以漳州特色花卉文化为主题,通过地面雕刻、雕塑小品、浮雕、绿植等,营造充满生活气息和文化底蕴的花巷空间,并开创性地将闽南红砖墙与照壁结合,打造独一无二的特色入户空间。

LANDSCAPE DESIGN

The landscape design pays great attention to Nine-Li Southern Fujian and Flowery Ancient Alley. By linking together the entrance plaza, community gate, flowery ancient alley, public neighborhood center, private courtyard and other marching sequences, the design successfully creates the sweet neighborhood space atmosphere. By introducing floor carving, sculpture accessories, reliefs and plants under the theme of Featured Zhangzhou Flower Culture, the design presents a flowery alley space imbued with life flavor and culture connotation. By combining red brick wall with screen wall for the first time, the design delivers a unique entry space.

INTERNATIONAL SHOPPING PLAZA CONCEPT COMPETITION

「概念商业广场」国际设计竞赛

INTERNATIONAL SHOPPING PLAZA CONCEPT COMPETITION ORGANIZING COMMITTEE

引领业界的创举——万达首届"概念商业广场"国际设计竞赛

| "概念商业广场"国际竞赛组委会评委、万达商业地产股份有限公司高级副总裁　赖建燕

"概念"一词是与热点、时尚、科技智能、理性与感性综合以及创新等密切关联的。概念车、概念手机等概念产品，已经公认地成为该领域产品未来突破的方向标。"概念商业广场"的范畴更为复杂，不单涉及建筑学、社会学、商业与金融领域，也包括了当今领先或迅速应用的智能科技。"概念商业广场"国际竞赛是对未来商业的探索，是对商业广场的物理空间与人的活动行为的研讨。

"概念商业广场"国际竞赛的构想，首先得到了同济大学设计院丁洁民院长的积极推动，使我们能够与中国建筑文化研究会共同主办、与同济大学建筑与规划学院联合主办这一竞赛。"概念商业广场"竞赛这一理念一经推出，就得到了哈佛大学设计研究生院、牛津大学、剑桥大学、英国建筑联盟学院、清华大学、东南大学、澳大利亚悉尼大学、澳大利亚墨尔本理工大学、新西兰奥克兰大学等国内外 35 家高校，以及法国建筑师委员会、意大利建筑师协会、以色列建筑师协会的响应支持。同时，我们非常荣幸地邀请到国际著名建筑设计大师丹尼尔·里伯斯金和建筑教育家伊纳克·阿巴罗斯分别担任本次竞赛专业组、学生组评审委员会主席，也邀请到第 77 届奥斯卡特效导演奖获得者安东尼·拉默里纳拉作为本次竞赛的顾问。

关于本次竞赛的成果形式，我与丹尼尔·里伯斯金先生进行了充分的交流。组委会内部进行了激烈的辩论，最终摒弃已经确定的传统的成果方案，采用了里伯斯金先生建议的不限成果的作品方案。这一方案得到了本次评委的认同，使竞赛形式本身即成为一种创新！本次竞赛的组织、议程、形式等，得到了中国建筑文化研究会刘凌宏副会长、北京市建筑设计研究院朱小地董事长、同济大学建筑与规划学院李振宇院长及李麟学教授、UED杂志主编彭礼孝、UED杂志执行主编柳青、万达商业规划院副院长风雪昆的诸多建议与支持。由于他们的推动，最终确定了竞赛与论坛两种形式共同推进，竞赛分为专业组与学生组并设两组评委的整体方案。考虑到中国商业广场起步较晚，两组评委的主席均由国际大师担任。

2015—2016"概念商业广场"国际建筑设计竞赛，是全球首个以"概念商业广场"为主题的不拘泥于形式的国际竞赛。竞赛自 2015 年 9 月 26 日发布以来，

The word "concept" is closely connected to hot spots, fashion, technology intelligence, comprehensive sense and sensibility, as well as innovation. Conceptual products such as concept cars and concept phones are developed to be a subject for a certain industry's future development, whereas concept commercial plaza presents a more complex field, which not only involves architecture, sociology, business and financial sectors, but also covers today's most advanced intelligent technology which is applied fastest. The International Competition of Concept Commercial Plaza attempts to explore future business and investigate the physical space of commercial plaza and people's activities and behaviors.

The idea of "International Shopping Plaza Concept Competition" is actively echoed by Mr. Ding Jieming, Director of Tongji University Architectural Design (Group) Co., Ltd, which enables us to co-host this competition with Architecture and Culture Society of China, College of Architecture and Urban Planning, Tongji University. The idea of concept commercial plaza competition also receives a positive response and support from Harvard Graduate School of Design, University of Oxford, University of Cambridge, Architectural Association School of Architecture, Tsinghua University, Southeast University, the University of Sydney, RMIT University, the University of Auckland, and other 35 universities at home and abroad along with professional bodies such as Conseil National De L'Ordre Des Architectes (CNOA) and Consiglio Nazionale Architetti Pianificatori Paesaggisti Conservatori (CNAPPC) and Israel Institute of Architects. Meanwhile, we are very pleased to invite internationally renowned architect Mr. Daniel Libeskind as the jury president for professional group and architectural educator Inaki Abalos for the students group. Anthony LaMoninara (best effects director of the 77th Oscar Award) is also invited as a consultant of this competition.

In terms of the form of achievements, Daniel and I had sufficient communication. The organizing committee also carried out passionate debates and ultimately decided to abandon the traditional fixed format and adopt the format with no limit to deliverables, which was initially suggested by Daniel. As this format was recognized by committee members, the competition became an innovation! This competition received considerable support on organizing, programming and forming from Mr. Liu Linghong, vice chairman of Architecture and Culture Society of China; Mr. Zhu Xiaodi, chairman of the board, Beijing Institute of Architectural Design; Mr. Li Zhenyu, Dean of College of Architecture and Urban Planning, Tongji University; Professor Li Linxue of College of Architecture and Urban Planning, Tongji University; Peng Lixiao, Chief editor of UED magazine; Ms. Liu Qing, Executive editor of UED magazine; Mr. Feng Xuekun, Vice President of Wanda Commercial Planning and Research Institute. Thanks to their efforts, two parallel programmes, competition and forum go hand in hand. There are also two categories in the competition: the Student Group and the Professional Group with two judge groups respectively. Considering that commercial plazas have a late start in China, the two judge groups are chaired by two world renowned masters.

截至2016年3月10日,参赛者遍及五大洲,报名团队达到1746组,最终有效交稿作品312份,使得本竞赛虽然首次举办就成为行业内覆盖面广、参与度高的全球性商业设计竞赛。竞赛成果中有8个为视频作品,虽然没有出现诗歌、戏剧等其他非建筑艺术作品,但这些视频作品却给评委和受众带来了强烈的冲击。竞赛作品评审于2016年3月26日、27日在万达学院举行。评审采用国际竞赛通用的三轮淘汰晋级评审方法:第一轮由各评委投出非晋级作品,对无争议的淘汰,对有争议的由主席组织讨论是否进入下一轮,第二、三轮由各评委投出晋级作品。三轮评比后,由评委会主席组织对争议作品进行讨论,最终由主席决定胜出作品,确保评审的公正性。通过两天的评审,专业组和学生组分别评选出一等奖一组、二等奖两组、三等奖三组以及优秀奖10组共百余幅获奖作品,竞赛取得了非常丰硕的成果(图1、图2)。

中国正处在城镇化进程的快速转变期,中国的商业模式也处在活跃的转型阶段。竞赛期间,组委会同时安排了"概念商业广场"主题研讨,专门举办了题为"破土"、"从物理空间到行为模式的探索"、"商业改变城市"、"商业广场中的体验与互动"、"商业广场中的艺术探索"与"为儿童而设计"六场"主题沙龙"以及"商业改变城市"高峰论坛,参与嘉宾包括城市决策者、管理者、国内外知名建筑师、学者、开发商、艺术家、商人、金融家、媒体人、互联网专家等百余位专业人士,真正做到了跨行业的深度交流,为未来商业模式提供了丰富的资讯!

本次竞赛得到了国内外上百家有影响力的专业媒体的推介和报道。欧洲、北美、澳洲等近50家境外专业媒体(如DEZEEN, Architects Journal, Architectural Review等),以及国内多家网络媒体(如新华网、凤凰网、人民网、中华建设等)对竞赛过行了详细报道。北京设计周组织者曾辉先生、原比利时驻华大使夫人顾菁女士、万达商业地产境外地产中心刘拥军副总经理、万达商业规划院副总规划师方芳女士在此项工作的组织中给予了出色的支持。

商业广场是城市美好生活的重要组成部分。概念物化创新——"概念商业广场"设计竞赛是基于建筑学与社会学"双重领域"交互的尝试性探索。我希望所有组织者、参赛者和参与者能在这个过程中收获乐趣与智慧,将"概念"作为与未来对话的沉淀,探索一种新的生活方式和一种新的生活理念,这将是我们持续举办这一竞赛的目的。

International Shopping Plaza Concept Competition 2015-2016 is the world's first commercial plaza themed competition and all the formalities were set aside. The competition was announced on 26, September, 2015. Till its end on 10 March, 2016, there were 1746 registration teams across 5 continents, and 312 valid works were received. Although it was held for the first time, this contest has become a global commercial design competition, making the greatest industrial coverage with a high degree of participation! There were eight video works amongst all the submissions. Although no other non-architecture art works, such as poetry or opera, were collected, those video works still brought mind-blowing refreshment to judges and audience. All the competition works were reviewed from 27 to 28 March 2016, at Wanda College. The process adopted an international three round elimination procedure. In the first round, the judges voted for "the eliminated". Then unquestioned works were eliminated while debatable works would be discussed until an agreement was reached. In the second round and the third round, judges voted for the "the remained". After the three-round procedure, the chairman for each group held a meeting to discuss about those debatable works and had the final right of decisions for any controversial works to ensure the fairness and objectiveness of the review. After two days of careful selection, professional and student groups each selected one first prize, two second prizes, three third prizes, and ten awards of excellence; it was indeed a very fruitful result!

China's urbanization develops fast during its transformation, and domestic business model is also in active transition phase. During the competition, the organizing committee held series of forums about concept commercial plaza, and especially held six theme forums such as "ground breaking", "exploration from physical space to behavioral pattern", "commerce changes cities", "experience and interaction in the commercial plaza", "artistic exploration in the commercial plaza" and "design for children" as well as "commerce changes cities" summit forum. The speakers and guests include urban policy makers, managers, national and international architects and scholars, developers, artists, businessmen and financiers, press, internet specialists and over one hundred professionals, achieving comprehensive communication across industrial borders and shedding light on the future business development with abundant information!

This competition has also gained publicity from hundreds of professional media organizations at home and abroad. Over 50 foreign media in Europe, North America and Australia, such as DEZEEN, Architects Journal, Architectural Review, and domestic media such as xinhuanet.com, ifeng. com, people.com.cn China Construction carried full reports about this competition. Mr. Zeng Hui, the organizer of Beijing Design Week, Ms. Gu Jing, spouse of Former Belgian Ambassador to China, Mr. Liu Yongjun, general manager of Wanda international real estate centre, Ms. Fang Fang, Deputy chief planner of Wanda Commercial Research and Planning Institute have all provided outstanding assistance to the competition.

Commercial Shopping Plaza is an important part of the ideal city life. Concept embodies innovation, concept realizes innovation - the competition is the tentative exploration based on the interaction between architecture and sociology. I hope that all the organizers, competition participants and staff will benefit from this process and harvest joy and wisdom, and everyone could consider "Concept" as a conclusion of a conversation with the future, exploring a new lifestyle and a new life concept.

图1 评审现场

图2 获奖作品

FUNCTIONAL GROUP AWARD
专业组获奖作品

FRIST PRIZE: USB URBAN SHOPPING BRIDGE
一等奖：USB URBAN SHOPPING BRIDGE

COMPANY PLUSOUT DESIGN STUDIO
单位 PLUSOUT 设计工作室
AUTHOR Musmeci Santi Maccarrone Sebastiano Shi Hongmei Ardizzi Andrea Sebastiani Alessandra
作者 Musmeci Santi Maccarrone Sebastiano 史洪美 Ardizzi Andrea Sebastiani Alessandra

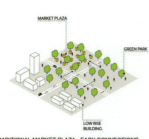
TRADITIONAL MARKET PLAZA - EASY CONNECTIONS

EVOLUTION OF THE CITY - LIMITED CONNECTIONS

MODERN CITY - LOSS CONNECTIONS AND IDENTITY

MODERN CITY - CREATION OF PLATFORM BRIDGE

FUTURE CITY - SHOPPING MALL CREATES CONNECTION WITHIN THE CITY

Through history the idea of SHOPPING has seen several changes. From the ancient form of the Roman market, where the architectural complex faced the plaza for events and activities of any kind, to the linear form of the street shops of the medieval times, afterwards covered with arcades.
In the 21st century the places we shop have lost the idea of mixing PUBLIC INTERACTIONS, DESIRE AND COMMERCE which were the foundations of the past.
In the reality of the contemporary city, where there is the pauperization of public spaces and green islands, the shopping mall must find its original qualities of meeting place and take back its role of one of the key space of urban planning.
But in a densely populated contemporary city, where is the possibility to design a PUBLIC MEETING SPACE? By definition the public space aims to re-connect points of the urban fabric otherwise isolated by the growth of the city's infrastructure.
It is in these broken points that our intervention aims to RECONNECT. The entire concept of a "shopping mall" is then integrated with the idea of being a "place of transition", a crossing point. This brings with it a transformation of its physical space.
The shopping mall of the future is no more served by infrastructure, but it itself becomes THE INFRASTRUCTURE. It takes back all the original qualities of the market plaza becoming a SHOPPING PLAZA where the main characters are the users and their activities.

SECOND PRIZES 二等奖: DISPLAY DEVICE

AUTHOR　　Yang Baoguang
作者　　　杨宝光

SECOND PRIZES 二等奖：NO MALL

COMPANY　CCTEG CHONGQING ENGINEERING CO.LTD.THE 4TH ARCHITECTURAL DESIGN INTITUTE
单位　　中煤科工集团重庆设计研究院有限公司第四建筑设计院
AUTHOR　Li Yingjun　Lu Linghui　Liu Junji　Ran Xiaoxia　Wu Lei　Wu Wei
作者　　李英军　卢凌晖　李平　刘俊吉　冉晓夏　吴磊　吴畏

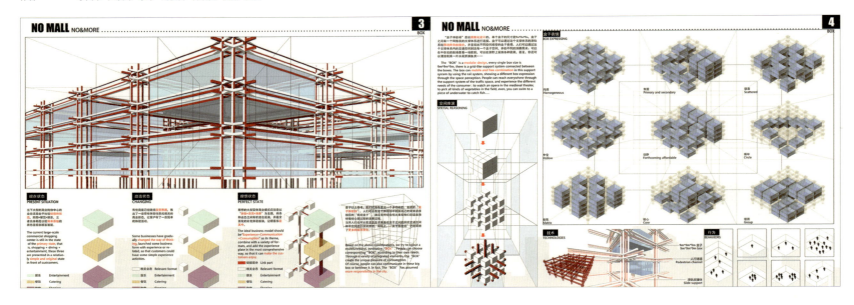

THIRD PRIZES 三等奖：CRUISE 漫游

COMPANY　GDF ARCHITECTURE(HK) CO.,LTD.
单位　　华凯国际（香港）有限公司
AUTHOR　Gao Shanxing
作者　　高山兴

THIRD PRIZES 三等奖：FLOATING CARVINAL 漂浮嘉年华

COMPANY　BEIJING VICTORY STAR ARCHITECTURAL & CIVIL ENIGEERING DESIGN CO.LTD.
单位　　北京维拓时代建筑设计股份有限公司
AUTHOR　Lei Ting　Wang Dayu　Zhao Bing　Li Yuanyuan
作者　　雷霆　王大宇　赵冰　李媛媛

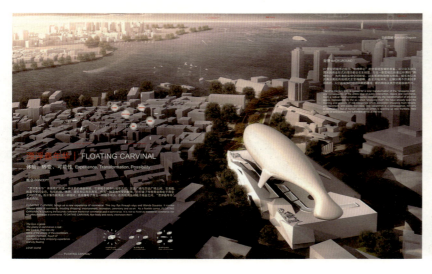

MERIT AWARDS 优秀奖

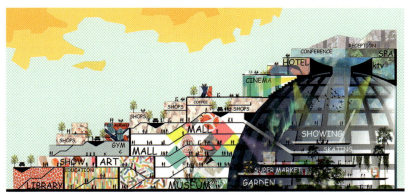

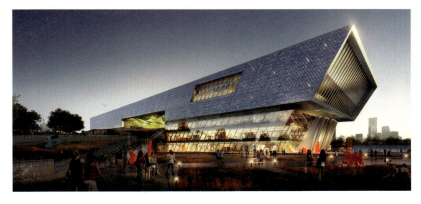

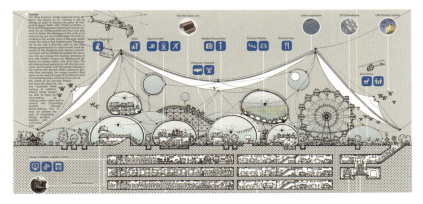

STUDENT GROUP AWARD
学生组获奖作品

FRIST PRIZE 一等奖 : THE CALL FROM THE WILD
野性的呼唤——诗意抵抗城市

SCHOOL	SCHOOL OF ARCHITECTURE, HARBIN INSTITUTE OF TECHNOLOGY
	SCHOOL OF ARCHITECTURE, SOUTHEAST UNIVERSITY
学校	哈尔滨工业大学建筑学院 东南大学建筑学院
AUTHOR	Luo Zhaoyang Wei Tangchenxi
作者	骆肇阳 魏唐辰希

SECOND PRIZES 二等奖：
MAKERS MARKET MADRID—COMMERCIAL, PRODUCTIVE AND PUBLIC INFRASTRUCTURE IN MADRID RÍO

SCHOOL	TECHNICAL SCHOOL OF ARCHITECTURE OF MADRID(ETSAM)
学校	马德里理工大学建筑学院
AUTHOR	Noemí Gómez Lobo
作者	Noemí Gómez Lobo

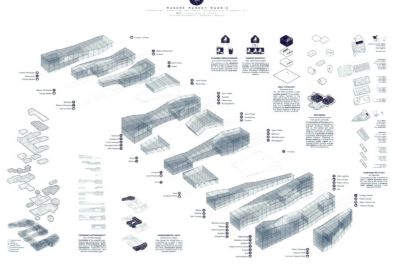

SECOND PRIZES 二等奖：CITY TREE OF DREAM

SCHOOL	XI'AN UNIVERSITY OF ARCHITECTURE AND TECHNOLOGY
学校	西安建筑科技大学
AUTHOR	Shen Yijun Hu Yihong Wu Mufei Liu Yueyi
作者	沈逸君 胡已宏 吴慕飞 刘悦怡

THIRD PRIZES 三等奖：NEW SILK ROAD

SCHOOL 学校	ILNAR AKHTIAMOV, RESEDA AKHTIAMOVA FACULTY OF ARCHITECTURE AND URBAN PLANNING, JILIN JIANZHU UNIVERSITY
	ILNAR AKHTIAMOV, RESEDA AKHTIAMOVA 吉林建筑大学建筑与规划学院
AUTHOR 作者	ANNA ANDRONOVA SUN LIDONG
	Anna Andronova 孙立东

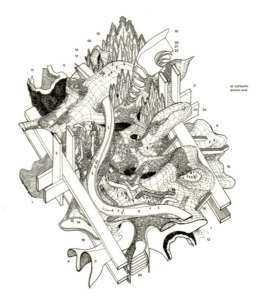
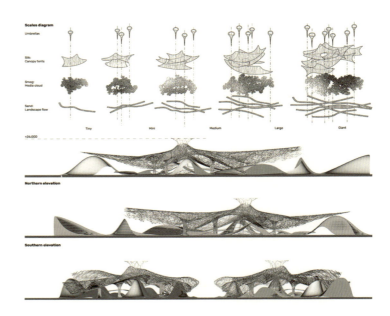

THIRD PRIZES 三等奖：ZIPCOMM 市城 城市

SCHOOL 学校	SCHOOL OF ARCHITECTURE, SOUTHEAST UNIVERSITY
	东南大学建筑学院
AUTHOR 作者	Lu Bingyang Fang Dongqing Sun Yi
	陆冰洋 方冬清 孙毅

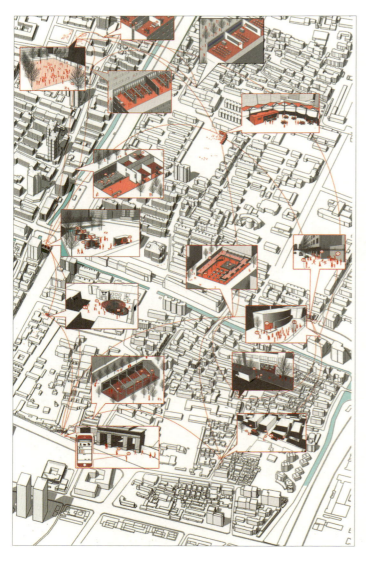

THIRD PRIZES 三等奖：THE URBAN ENGINE 城市动力

SCHOOL 学校	TIANJIN UNIVERSITY
	天津大学
AUTHOR 作者	Wang Liyang
	王立杨

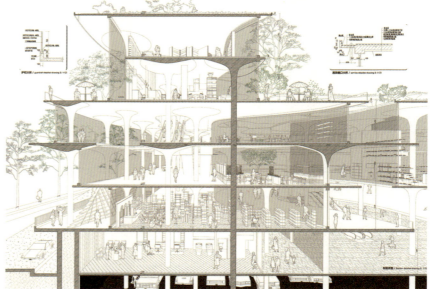

MERIT AWARDS 优秀奖

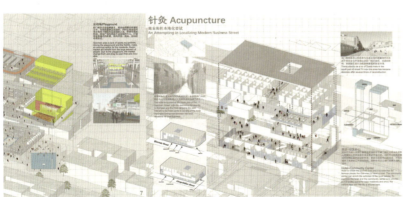
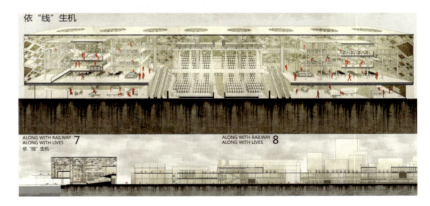

FORUM 论坛

2016 年 3 月 29 日，由中国建筑文化研究会、万达商业规划研究院主办、同济大学建筑与城市规划学院、同济大学建筑设计研究院(集团)有限公司联合主办的"商业改变城市——'概念商业广场'国际建筑设计竞赛高峰论坛"于中华圣公会教堂成功举行。本次论坛缘起于"概念商业广场"国际建筑设计竞赛，以"商业·城市—— 商业设计的城市探讨、商业·跨界——商业设计的多元探索"等热点为主题，深入探讨未来人与城市、人与行为、人与商业的关系。

概念商业广场竞赛的理念一经推出就受到了各界人士的广泛关注，也由此引发了系列论坛的成功举办与我们更加深入的思考：中国的商业模式正处于高速转型期，商业广场则是城市美好生活的重要组成部分。这一竞赛是一种基于建筑学和社会学双重领域交互的尝试性的探索：有之以为利，无之以为用，主张竞赛与论坛同等重要，理想和现实同等重要，设计类成果和非设计类成果同等重要。基于概念商业广场竞赛的系列活动将持续展开。

On March 29th, 2016, Commerce Changes Cities· International Shopping Plaza Concept Competition Summit Forum was held. The forum was organized by Architecture and Culture Society of China and Wanda Commercial Planning & Research Institute and co-organized by College of Architecture and Urban Planning, Tongji University. Originated from the International Shopping Plaza Concept Competition, this forum focused around heated topics such as "Commerce · Cities—The Urban Issues of Commercial Designs" and "Commerce · Crossover—The Pluralistic Exploration of Commercial Designs" to invoke in depth discussions about the relationship between man and city, man and behavior, as well as man and business in the future.

Since its initiation, the competition concept of "Concept Commercial Plaza" has been a widespread public topic, which has triggered a series of successfully held forums and our deeper thinking: China's business model is in a high-speed transition period, and an important part of wonderful urban life is commercial plazas. The competition is a tentative exploration based on the interaction of architecture and sociology: What is useful will do good, and what seems useless may become useful. The competition is as important as the forum, ideals are as important as realities, and design achievements are as important as non-design achievements. A series of events based on the concept commercial plaza competition will continue to take place.

PART F INTERNATIONAL SHOPPING PLAZA CONCEPT COMPETITION
"概念商业广场"国际设计竞赛

APPRAISAL
评奖

AWARD CEREMONY
颁奖

DESIGN & CONTROL
设计及管控

R&D EVENTS OF WANDA BIM GENERAL CONTRACTING MANAGEMENT
"万达 BIM 总发包管理模式"研发大事记

万达商业地产高级总裁助理兼设计中心总经理、技术研发部总经理　尹强
万达商业地产技术研发部副总经理　朱镇北

2015年8月，在万达集团丁本锡总裁直接领导下，"万达BIM研发小组"联合"设计总包"、"工程总包"和"工程监理""四方"，启动"BIM总发包管理模式"研发工作。BIM研发工作历时17个月，经历如下10个重要的环节。

1. 2015年8月17日，成立"BIM总发包管理模式"工作组。
- 集团丁本锡总裁听取"BIM总发包管理模式"思路汇报；
- 确定"BIM总发包管理模式"研发方向（图1）；
- 成立"BIM总发包管理模式"集团研发领导小组；
- 确定"四方"建立"BIM总发包管理模式"研发工作组，确定BIM研发构想。

2. 2015年11月20日，集团批准"BIM总发包管理模式"方案计划。
- 集团王健林董事长听取"BIM总发包管理模式"汇报；
- 集团正式批准"BIM总发包管理模式"研发方案及实施计划（图2）。

3. 2016年2月18日，召开"BIM总发包管理模式"研发启动大会。
- 确定年度"四方"培训计划；
- 安排"四方"参与"BIM总发包管理模式"研发。

4. 2016年6月30日，完成"四方标准"统一。
- 完成"四方"的"BIM总发包管理模式"技术标准；
- 统一"BIM总发包管理模式"的"四方"管理标准；
- 完成万达对应制度修编工作。

5. 2016年8月30日，完成"四方"《万达BIM总发包管理模式操作手册》。
- 完成"四方"《万达BIM总发包管理模式操作手册》；
- 确定"'四方'BIM总发包管理模式'四阶段'"计划模块。

6. 2016年9月30日，完成"BIM总发包管理平台"的研发。
- "BIM总发包管理平台"上线（图3）。

7. 2016年11月15日，完成BIM模型插件的研发。
- 完成BIM标准模型；

In Aug. 2015, under direct leadership of Ding Benxi, President of Wanda Group, four parties (i.e. Wanda BIM R&D Team, main design contractor, main project contractor and supervisor) launched the Development work of BIM General Contracting Management (BIM Management). The 17-month long R&D work has undergone the following ten steps.

1. ON AUG. 17, 2015, BIM MANAGEMENT WORK GROUP WAS ESTABLISHED.
- President Ding Benxi reviewed to the report on BIM Management strategy;
- R&D direction of BIM Management was determined (Fig.1);
- Wanda Group R&D leading group for BIM Management was set up;
- A Four-Party BIM Management R&D Work Group was organized for joint research.

2. ON NOV. 20, 2015, WANDA GROUP APPROVED THE BIM MANAGEMENT SCHEME AND PLAN.
- Chairman Wang Jianlin reviewed to the report on BIM Management;
- Wanda Group officially approved R&D scheme and implementation plan for BIM Management (Fig.2).

3. ON FEB. 18, 2016, BIM MANAGEMENT R&D KICK-OFF MEETING WAS CONVENED.
- Annual Four-Party Training plan was determined;
- Four parties participated BIM Management R&D together.

4. ON JUNE 30, 2016, FOUR-PARTY UNIFIED STANDARD WAS COMPLETED.
- Four-party technical standards of BIM Management were finished;
- Four-party management standards of BIM Management were unified;
- Editing and revision work for corresponding Wanda system was completed.

5. ON AUG. 30, 2016, FOUR-PARTY *OPERATION MANUAL OF WANDA BIM GENERAL CONTRACTING MANAGEMENT* WAS COMPLETED.
- Four-party *Operation Manual of Wanda BIM General Contracting Management* was completed;
- Four-stage planning modules of four-party BIM Management were determined.

6. ON SEPT. 30, 2016, BIM MANAGEMENT PLATFORM R&D WAS COMPLETED.

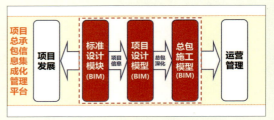

图1 BIM研发构想

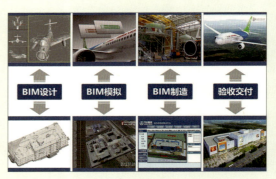

图2 BIM研发方案

图3 BIM总发包管理平台

- 完成BIM五大类功能插件；
- 确定"BIM总发包管理模式"试点项目。

8. 2016年11月30日，完成"BIM总发包管理模式"集中模拟测试。
- 完成两次"四方"模拟测试及复检；
- 完成"BIM总发包管理模式"构件库研发；
- 完成"BIM总发包管理模式"标准材料库研发。

9. 2016年12月26日，完成"BIM总发包管理模式"复盘总结。
- 集团丁本锡总裁听取汇报；
- 集团批准2017年"BIM总发包管理模式"计划及试运行计划。

10. 2017年1月9日，召开"BIM总发包管理模式"及"慧云3.0版"运行启动大会（图4、图5）。

"万达BIM总发包管理模式"研发工作历时17个月，按计划完成"四大类"研发成果，开创性应用"测试——试点——复盘——培训"的"四步骤工作法"，采用"边研发、边培训"的推进模式，于2017年1月1日开始在万达直投项目中"试运行"。

- Project management platform of BIM Management went live (Fig.3).

7. ON NOV. 15, 2016, R&D OF BIM MANAGEMENT MODEL PLUG-INS WERE COMPLETED.
- Standard BIM model was finished;
- Five kinds of functional plug-ins for BIM Management were completed;
- BIM Management pilot project was initiated.

8. ON NOV. 30, 2016, CENTRALIZED SIMULATION TEST FOR BIM MANAGEMENT WAS COMPLETED.
- Two rounds of Four-party simulation tests and retests were conducted;
- BIM Management object Library was finished;
- BIM Management Standard Material Library was finished.

9. ON DEC. 26, 2016, BIM MANAGEMENT CHECK & SUMMARY WAS COMPLETED.
- President Ding Benxi reviewed progress report;
- Wanda Group approved overall schedule and trial run plan of BIM Management for 2017.

10. ON JAN. 9, 2017, BIM MANAGEMENT AND "HUIYUN V3.0" OPERATION KICK-OFF MEETING WAS CONVENED (Fig4, Fig.5).

After 17 months R&D work for BIM Management four-type of results has achieved, innovatively applied four-step working method (test-pilot-check-training) and adopted progression model of researching while training. On Jan. 1, 2017, BIM Management started trial run on Wanda's directly invested projects.

图4 BIM总发包管理及慧云3.0版运行启动大会

图5 BIM四大核心成果

SUMMARY FOR R&D RESULTS OF WANDA BIM GENERAL CONTRACTING MANAGEMENT IN 2016
2016"万达 BIM 总发包管理模式"研发成果总结

万达商业地产高级总裁助理兼设计中心总经理
技术研发部总经理　尹强

2016年，万达商业地产技术研发部会同万达商业地产设计中心、万达商业地产计划管理部、万达商业地产成本控制部、万达商业地产质量监管中心及万达集团信息中心等相关研发单位，超额50%（计划研发立项11项，实际完成18项）完成全"万达BIM总发包管理模式"研发工作。

"BIM总发包管理模式"研发工作包括"四大类成果"：模型、插件、制度、平台。技术研发部主要负责其中标准版模型、构件库标准、插件的研发工作，具体成果如下：

一、标准版模型

标准版模型在3D模型基础上输入数字信息，实现与成本、计划和质量管理需求的信息挂接。万达"BIM总发包管理模式"模型为满足各管理部门的管理需求，在模型信息和模型深度方面均提高了相关要求：万达BIM模型可达到LOD400~LOD500深度，而常规设计模型仅达到LOD300深度。

1. 设计信息
模型涵盖12个分项专业的全专业信息，录入标准材料库编码、材质贴图与标准材料库一一对应（图1）。

In 2016, R&D dept. of Wanda Commercial Planning and Research Institute (WCPRI) together with other business units such as Design Center, Planning Management Dept., Cost Control Dept. and Quality Supervision Center at Wanda Commercial Properties and Information Center of Wanda Group has over fulfilled the R&D tasks of Wanda BIM General Contracting Management (BIM Management) for an excess of 50% (planned R&D projects: 11; actual projects: 18).

BIM Management R&D results are divided into four types: (1) BIM Model; (2)Design plug-ins; (3)management policies; (4) management platforms. while R&D dept. organized below technology works including standard models object library. BIM standards & plug-ins.

I. STANDARD MODEL

Standard models add cost information, schedule information and quality management demand information to 3D models. BIM Management models boast of improved information and level of details (LOD400-LOD500 compared to LOD300 available to the conventional design models), so as to cater for management demands of diverse departments.

1. DESIGN INFORMATION
Models have covered full-discipline (12 disciplines) information, and have standard material library code and texture in perfect mapping with standard material library (Fig.1).

图1 模型设计属性页面

图2 模型成本属性页面

图3 模型计划属性页面

图4 模型质量属性页面

2.成本信息

模型构件与成本清单挂接，通过插件实现一键算量（图2）。

3.计划信息

模型构件与模块化节点挂接，利用BIM模型进行计划管理，确保计划可控（图3）。

4.质量信息

模型构件与质量验收标准挂接，现场验收时，可以实时查阅质量标准，实现标准统一（图4）。

万达BIM标准版模型作为BIM研发的重要成果之一，将与二维图纸成为现场施工的重要依据，是进行过程管控的工具，是"四方"（万达、设计总包、工程总包和工程监理）信息交互的基础。

二、构件（族）库

万达构件（族）库，是国内第一个成体系的企业级文件合集，包括12类1.1万个标准构件。每个标准构件挂接6D业务信息并实现参数化设计，已涵盖全部业

2. COST INFORMATION

Model component is linked with cost list. With help of plug-ins, amount is calculated with one click (Fig.2).

3. SCHEDULE INFORMATION

Model component is connected with modular node. schedule is managed via BIM to ensure it is controllable (Fig.3).

4. QUALITY INFORMATION

Model component is linked with quality acceptance standard. During site acceptance, quality standard is reviewed in a real-time manner to enable it to be unified (Fig.4).

Being one of the important results BIM R&D has achieved, BIM standard model, together with 2D drawings, serves as an important basis for on-site construction and a tool for process control. It constitutes the foundation of information exchange among Four Parties (i.e. Wanda Group, main design contractor, main project contractor and project supervisor).

II. OBJECT (FAMILY) LIBRARY

Wanda object (family) library is the first systematic enterprise-level document collection in China, including 11,000 standard components in 12 categories. With

图5 万达构件库

务需求，关联已有效果类材料库、设计标准图集和成本计价信息（图5）。

三、系列标准

万达"BIM总发包管理模式"的"三大系列"、"九个标准"是全球首创的贯穿商业地产项目全生命期BIM应用的系统性企业标准。参照国家BIM标准体系，万达BIM系列标准包括三个层级：第一层级为标准总则；第二层级为平台建设系列标准、模型系列标准、数据信息系列标准；第三层级为与标准相对应的指导具体工作行为和技术细节的规范、审查要点、清单、指南、用户手册等。

万达"BIM系列标准"和"BIM建模指南"经过大量建设项目检验和修正升级，对其他企业乃至全行业制订可实施、可操作的BIM标准都具有参考借鉴价值。

四、功能插件

功能插件是利用BIM技术开发的自动化、信息化辅助工具。2016年已完成构件库插件、编码插件、材料设备插件、模型检查插件共"四大类功能"。

1.构件库插件

辅助构件的设计和布置，在BIM模型中自动形成构件，通过下拉菜单确定构件信息，自动验证构件"合规性"，真正实现标准固化（图6）。

linked 6D business information and parameterized design, standard object have already covered all business requirements and related to the existing effect material library, design standard atlas and cost pricing information (Fig.5).

III. BIM SERIES OF STANDARDS

The "Three Series" and "Nine Standards" initiated by BIM Management is the world's first systemic enterprise standard that runs through the full-process BIM application of commercial properties projects. With reference to the national BIM standard system, Wanda BIM series and standards include three levels: first level refers to general provisions of standards; second level touches standards of platform construction series, model series and data information series; third level involves codes, review points, BoQs, guidelines and user manuals that guide work practice and technical details in line with standards.

Being tested by a considerable amount of construction projects and adjusted and upgraded, Wanda BIM Series of BIM Standard and BIM Modeling Guide have a great value for other enterprises and even the industry in promulgating enforceable and operable BIM standard.

IV. FUNCTIONAL PLUG-INS

Functional Plug-ins are an automated and informationized auxiliary tool developed on BIM technology. In 2016, four kinds of functional plug-ins have completed, including component library plug-in, coding plug-in, material & equipment plug-in and model checking plug-in.

1. COMPONENT PLUG-IN

It assists design and layout of components, automatically

图6 构件库插件

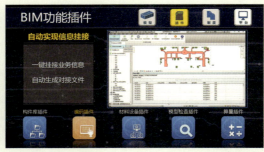

图7 编码插件

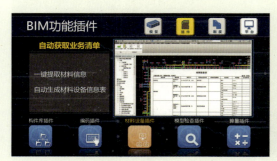

图8 材料设备插件

图9 模型检查插件

2. 编码插件

实现BIM构件与业务信息自动化对应与匹配，是模型标准的信息化、智能化的保障，是BIM模型与其他专业软件、系统、平台对接的基础（图7）。

3. 材料设备插件

实现了自动获取业务清单，一键提取材料信息，自动生成材料设备信息表（图8）。

4. 模型检查插件

可以自动化检查模型质量，自动比对图纸差异，自动判定项目类型，减少人工检查工作量（图9）。

五、结语

2016年万达"BIM总发包管理模式"研发项目全部按计划圆满完成。研发管理采用多方集中测试、联合办公、项目试点、复盘总结、战略合作、建立专家咨询库、知识产权保护全覆盖、多维度同步培训等创新研发模式，为"BIM总发包管理模式"研发成果的落地奠定了坚实的基础。

build up objects in BIM models, determines object information through pull-down menu, and verifies compliance of objects automatically. Standard is thus virtually specified (Fig.6).

2. CODING PLUG-IN

It enables automatic correspondence and match between BIM components and business information, safeguarding informationization and intellectualization of model standards and serving as the basis for BIM models to connect with other professional software, systems and platforms (Fig.7).

3. MATERIAL & EQUIPMENT PLUG-IN

It can automatically acquire business tasks list, extract material information with one click to automatically generate material & equipment list (Fig.8).

4. MODEL CHECKING PLUG-IN

It checks model quality, compares drawings and determines project type all in automatic manner, reducing manual check workload (Fig.9).

V. CONCLUSION

In 2016, BIM Management R&D projects have been successfully completed as scheduled. R&D management innovatively adopts multiple models to lay a solid foundation for implementing BIM Management Model, including multi-party centralized test, co-working, pilot project, check & summary, strategic cooperation, expert consultation library, complete coverage of intellectual property protection and multidimensional synchronous training.

SUMMARY FOR IMPLEMENTATION OF DESIGN GENERAL CONTRACTING MODEL

"设计总发包模式"推行小结

万达商业规划研究院副院长　张东光
万达商业规划研究院主任工程师　李万顺

为确保"设计总发包模式"推行成功，万达规划系统进行了缜密的安排和策划、集团各部门联动，最终确保"设计总发包模式"和"BIM总发包模式"在2016年年底同步研发完成，计划于2017年年初推行，关于设计总发包模式小结如下。

一、推行设计总发包模式的阶段

1.前期准备阶段（2015年）

（1）"设计总包"管理文件编制完成了《设计总包招标文件》、《设计总包、分包合同模板》、《设计总包操作手册》；（2）进行潜在设计总包单位的交流培训，让他们充分了解"总包"实施的办法和推行路径，并提出相关修改意见；（3）顺利建成三个销售物业试点项目，根据其过程中遇到的问题，编制和完善了持有物业设计"总包"的制度、流程、计划、权责等管理内容。

以上工作均已在2015年年底完成，具体可参考《万达商业规划2015——持有类物业》中的文章《"设计总发包模式"应用初探》。

2.试运行阶段（2016上半年）

（1）确定试点项目——根据摘牌时间，2016年年初选定四个持有物业项目作为设计总包试点项目。这四个项目分别为：江苏溧阳、湖南衡阳、浙江余杭和河南新乡。

（2）遴选"总包"单位——40多家万达优秀供方对4个试点项目进行"设计总包单位"的投标工作。经过评委层层筛选，最终表现优异的19家设计单位脱颖而出，作为该年度"设计总包"的重点单位；同时每个标段的第一名称为第一批实施试点项目的"设计总包单位"。

（3）强化培训和交流——2016全年共17家"设计总包单位"获得34个"持有物业设计总包"项目。"设计总包单位"择依据为综合实力、对分包控制力、对"总包"的理解力及地域优势；尤为重要的一个特点是：参与BIM标准版研发的"设计总包单位"异军突起，成为该年度获取项目的最大赢家。

I. STAGES OF IMPLEMENTING THE MODEL

1. PREPARATION STAGE (2015)

(1) Management documents of design general contracting have been compiled, including *Design General Contracting Bidding Document, Template of Design General Contracting and Sub-contracting Contracts and Design General Contracting Operation Manual*; (2) Communication and training have been provided for general design contractors, who can thus have a complete understanding of implementation measures and implementation paths for the Model and may provide valuable opinions for improvement; (3) Three pilot projects of properties for sale have been successfully established, and management contents including the system, flow, plan, rights and obligations for the design general contracting of properties for holding have been compiled and improved based on the problems encountered during the periods of the three pilot projects.

The work above has all been completed at the end of 2015, with details shown in *Discussion on Application of Design General Contracting Model* in Wanda Commercial Planning 2015: Properties for Holding.

2. TRIAL OPERATION STAGE (THE FIRST HALF OF 2016)

(1) Determining pilot projects: Four projects of the properties for holding were selected in early 2016 to be pilot projects for design general contracting, they were respectively the Liyang Project (Jiangsu Province), the Hengyang Project (Hunan Province), the Yuhang Project (Zhejiang Province) and the Xinxiang Project (Henan Province).

(2) Selecting general contracting companies: Among the 40 plus outstanding Wanda Suppliers bidding for general design contractors of the four pilot projects, 19 with excellent performance won the bidding after rounds of selection to be key design general contracting companies of the year; the top winner of each bid section is called general design contractor among the first batch of pilot projects.

(3) Stepping up training and communication: In the year 2016, 34 design general contracting projects of properties for holding were granted to a total of 17 general design contractors selected based on comprehensive strength, subcontracting control, understanding for subcontracting and regional advantage. It is particularly noted that the general design contractors engaged in R&D of standard version BIM have stood out to be the biggest winner of the year.

随着试行项目的全面实施,对"总包单位"的设计成果及现场协调管控也成为工作的重点。依据《设计总包操作手册》,2016年上半年对"设计总包"项目开始了"月度履约"评估考核,上半年为过渡期,履约成绩不计入全年成绩;7月份开始集团与项目公司设计部共同对"总包"进行月度考核打分,当月的成绩与排名并通过微信平台组建的"供方联盟"进行通报,单月的考核结果累积加权平均成绩将作为"设计总包单位"年度成绩。年度整体排名依据"操作手册供方履约评估管理办法"进行奖惩措施。优秀"总包单位"在2017年"时尚之春"供方大会进行颁奖,表现不好的单位将末位淘汰,所有考核结果都将在规划网站进行公示。

3. 全面运行阶段(2016下半年)

(1)四个试点项目复盘总结,完成管理文件终稿——2016年6月份,4个试点项目均完成了施工图设计,集团组织进行项目阶段性复盘。根据复盘结果,联合集团法务部、成本部及时梳理完成了设计总包及分包的合同文本、规范了合约体系,并在此基础上配合"BIM总发包模式"完成《BIM模式操作手册》。

(2)结合集团招采平台建设,完成设计总包和分包库的建立——结合集团成本部要求,2016年5月启动了万达广场及销售物业"设计供方库"的建立工作。"设计总包供方库"计划在2016年11月份完成,共约70余家设计单位报名参加"设计总包供方库",其中67家单位通过系统自动评审,经过综合评审及面试答辩等考核环节,最终19家"设计总包单位"入库,如图1所示。

为配合集团"BIM总发包管理模式"的推行,"设计总包单位"在入库评审条件中除了强调了企业团队实力、过往业绩外,还特别要求了BIM能力(包含:BIM团队考核、BIM技术考核、BIM建模考核),并在最终的答辩环节增设了BIM现场上机操作,为2017年"BIM设计总包"采用集中招标确定了基础。

为了确保"设计总包"的推行顺利,"分包库"的建立选择的方向与"总包单位"相同,待"设计总包"实施成熟后,将分步取消"分包库",由"总包单位"自行选择分包。

(3)年底"设计总包"年度复盘——为提升"设计总包"单位对项目管控的能力并顺利解决项目中出现的各种问题,2016年底组织"优秀总包"单位召开项目复盘交流会,针对项目管控的实际问题进行案例分享,将涉及制度管理文件内容进行修订。"设计总包"全年动态数据:"总包"及"分包"中标数量、年度履约成绩排名、设计变更、第三方审查、通报批评、通报表扬等全方位信息与"设计总包"单位进行通报。

图1 云系统3.0版运营维护简化

After full implementation of the pilot projects, the work focus shifts to control of design results and on-site coordination. According to *Design General Contracting Operation Manual*, the design general contracting projects are subject to monthly performance assessment in the first half of 2016, and the performance scores are not included into yearly score as the first half year is a transitional period. Since July, design general contracting has been monthly assessed jointly by the Group and project companies' design department, grades and rankings of the month are published on Suppliers Alliance available on Wechat. The aggregated weighted average score of assessment result in each single month is counted as yearly score of general design contractors. Incentives and disincentives are applied to yearly ranking as per Administrative Measures for Operation Manual's Suppliers Performance Assessment. Excellent main contractors will be awarded at the 2017 Spring of Fashion; companies with poor performance will be deleted from the list; assessment results will be made available on Wanda Planning System Website.

3. FULL OPERATION STAGE (THE SECOND HALF OF 2016)

(1) The four pilot projects were checked and summarized and final version of management documents were completed: In June 2016, construction drawings of the four pilot projects have been completed and phased check is applied by the Group. According to the check results, the Legal Department and Cost Department of the Group have organized and completed contract text of design general contracting and subcontract and have regulated contract system in a timely manner, based on which *BIM Model Operation Manual* has been finished in coordination with BIM General Contracting Model.

(2) Design general contractor and subcontractor list were established in combination with the Group's bidding and procurement platform construction: as required by the Group's Cost Department, the construction of Design Supplier List for Wanda Plazas and properties for sale has been started in May 2016, and is planned to be completed in November of the same year. In total, among the more than 70 designers which have signed up for the List, 67 have passed system automatic review and 19 have been enrolled after comprehensive review and interview & representation (Fig.1).

To assist in implementing the BIM General Contracting Model, general design contractors have also stressed on BIM capability (including BIM team assessment, BIM technical assessment, BIM modeling assessment) in the review, in addition to team strength and past performance of an enterprise, and have requested on-site hands-on BIM operation during representation. These efforts have paved the way for centralized bidding available to BIM Design General Contracting in 2017.

Subcontractor list has established in the same direction as that of main contractors, for a smooth advancement of design general contracting. After design general contracting is put on the right track, subcontractor list will be removed progressively and main contractors will select subcontractors on their own.

(3) Design General Contracting Annual Review: At the end of 2016, project check exchange meeting is convened among excellent main contractors, aiming to promote general design contractors' control capability in project control and to address problems found in project. At the meeting, the following work is done: practical problems facing project control have been shared, and documents concerning system management have been revised; yearly dynamic data of design general contracting has been revealed to

总结会上，年度优秀"总包"单位进行了管控经验的分析共享。万达也将2017年"BIM设计总包"推行的计划及管控的新需求进行讲解，让"总包"单位未雨绸缪，提前应对。

二、推行"设计总发包模式"的好处

经过2016年全年的"设计总包"推行，深刻体会到"上层建筑"的改革带来十分显著的效果。

1.节省人员，提升效率

（1）减少了招标数量——每个项目在"设计总包"推行之前，每个子项设计均需要进行单独招标，数量至少为8个，"设计总包"推行之后招标数量为1个。招标数量的减少，意味着招标文件编制，甲乙双方到场开标述标，组织评委评标、定标的工作内容都将大幅度减少。

（2）减少了协调会数量——每个项目在"设计总包"推行之前要开一百多个专业协调会。"设计总包"推行之后降到20个。甲方出差，会议组织次数减少。

"设计总包"推行符合集团"能外包尽量外包"的原则，减轻了万达设计管理人员的管理负担，进而减少人员数量，人员成本得以降低。"设计总包"单位增加了"设计总包"管理费，在增加收入的同时节省了投标成本、差旅成本等，最终实现了共赢，如图2所示。

2.计划可控、质量提升

"设计总包"承担着"总协调、总负责、总管理"的管理职责，在项目全过程中充分发挥了设计单位的主观能动性和积极性：根据整体项目推行计划进行三、四级的分解和细化，确保设计计划按时完成；承担起更多的技术协调，图纸中的"错漏碰缺"明显减少，大大提升了图纸质量，确保现场顺利施工（图3、图4）。

三、推行"设计总发包模式"的展望

2017年，万达集团将全面推行"BIM总发包模式"。此发包模式的前提是"设计总包"顺利推行。如何能够实现"带单发展、带单设计、带单建设、带单运营"，给"设计总包"单位提出更高的要求；摘牌后90天完成整套图纸及BIM模型是"设计总包"单位的巨大挑战。经过2016年的磨合，2017年"设计总包模式"将更加成熟、完善，通过众多的"设计总包"单位推广，此模式将引领行业方向和趋势。

执行力与创新力是万达企业文化的核心，转型之下，执行与创新是最好的"继往开来"，只有执行与创新才能解决问题！

general design contractors, including bid-winning quantity of main contractors and subcontractors, annual performance score ranking, design change, the third party review, notice of criticism and notice of commendation; excellent main contractors of the year have shared their experience in control; Wanda has expounded on it implementation plan for BIM Design General Contracting and new requirements on control in 2017, so that main contractors may take precautions.

II. BENEFITS OF IMPLEMENTING THE MODEL

After one year's implementation of the Model in 2016, the effect of the "top structure" reform is remarkable.

1. REDUCED PERSONNEL AND IMPROVED EFFICIENCY

(1) Reduced biddings: before implementing the Model, bidding for each subitem design needs to be separately conducted, totaling at least 8, while after implementing the Model, bidding is reduced to only one. The reduced biddings lead to considerable reduction in such work content as preparation of bidding documents, bid opening and presentation by Party A and Party B, organization of judges for bid evaluation and award of bid.

(2) Reduced coordination meetings: after implementing the Model, the coordination meetings by discipline are reduced to 20 from the previous more than one hundred ones. Business trips reduced and meetings are reduced accordingly.

The implementation of the Model conforms to the Group's principle of outsourcing as much as possible, which alleviates the management burden imposed on Wanda's design management personnel and further cuts down staff cost through reduced number of personnel. General design contractors, meanwhile, see increase in design general contracting management fee whereas decrease in bid cost and travel cost, and eventually achieved a win-win situation (Fig.2).

2. CONTROLLABLE PLAN AND PROMOTED QUALITY

Administratively, the Model takes charge of overall coordination, overall responsibility and overall management, so that designer may give full play to its initiative and enthusiasm in the whole process of projects: the model enables third- or fourth-level breakdown and subdivision of the overall project implementation plan, so as to ensure design is completed on time; the model stresses technical coordination, which greatly reduces "error, omission, collision and missing" in drawings to improve drawing quality and to safeguard smooth construction on site (Fig.3, Fig.4).

III. PROSPECTS OF IMPLEMENTING THE MODEL

In 2017, Wanda Group will put the BIM General Contracting Model in full operation, which must be based on successful implementation of design general contracting. General design contractors, however, are faced with a demanding requirement to make mission-oriented development, mission-oriented design and mission-oriented operation come true and with a huge challenge to complete the whole set of drawings and BIM models 90 days after delisting. After one year's practice, the Model will surely get more secure and perfect in 2017. By virtue of promotion by a multiple general design contractors, the Model will lead direction and trend of the industry.

Innovation and execution are the core of Wanda's culture. They are the key of exploring the future and also real solution of problems.

图2 设计总包资源整合优化提升效率

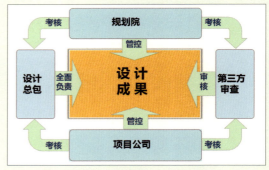

图3 总发包管理模式的设计成果管理体系

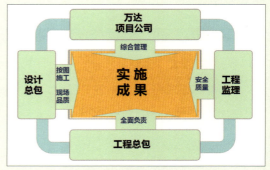

图4 总发包管理模式的现场质量管理体系

R&D AND APPLICATION OF HUIYUN SYSTEM V3.0
"慧云系统"（3.0版）的研发及应用

万达商业规划研究院副院长　方伟
万达商业规划研究院总工程师　范珑

"慧云系统"是万达集团自主研发、具有独立知识产权的智能化管理系统。2013年至2015年，"慧云系统"实现了由"1.0版"到"2.0版"的升级，2016年万达集团又研发了"慧云系统"3.0版（云平台版）其具有便于集中管控、利于大数据分析、易于简化维护三大特点。

Huiyun System is a kind of intelligent management system that has independent intellectual property right and is independently researched and developed by Wanda Group. From 2013 to 2015, Huiyun System has upgraded from V1.0 to V2.0; in 2016, Huiyun System V3.0 (cloud platform version) has been issued by Wanda Group. The new Huiyun System is characterized by being convenient for centralized control, big data analysis and maintenance.

一、"慧云系统"（3.0版）便于集中管控

"慧云"（3.0版）平台的发布，有助于集团总部集中管控"慧云"施工过程，确保实施质量，同时实现了软件开发标准化，所有广场统一使用总部"云版慧云系统"，版本统一。新版"慧云"产品部署在集团云数据中心，系统更加健壮和稳定。随着产品的标准化，可加强实施过程的标准化，更利于培养足够数量及质量的实施团队（图1）。

二、"慧云系统"（3.0版）利于大数据分析

"慧云"（3.0版）实现了总部"云端"实时获得现场数据并进行集中存储，为总部端实时监控打下基础，同时为后续大数据分析建立了基础的支撑。相对"慧云"（1.0版）、（2.0版）的单机部署方式，只有"云方案"的"慧云"（3.0版）所支持的高扩展、高弹性、集中运维的架构，才能满足"慧云"稳定运行及海量数据长期存储、分析的苛刻要求，并可支撑未来系统的发展演进（图2），系统并发请求最高可达3万次/秒，数据存储可达3千亿条/年。

I. CONVENIENT FOR CENTRALIZED CONTROL

After Huiyun V3.0 releases, the Group HQ (headquarters) finds it easier to centrally control Huiyun construction process and to safeguard implementation quality. Meanwhile, software development is standardized and all plazas are configured with the same version-the HQ cloud version Huiyun system. With the new version being deployed in the Group's cloud data center, the system gets more robust and stable. With the standardization of Huiyun products, the implementation process will be standardized as well, which further help cultivating implementation teams of sufficient quantity and quality, as shown in Fig. 1.

II. CONVENIENT FOR BIG DATA ANALYSIS

Thanks to Huiyun V3.0, HQ cloud can have the function of real-time data acquisition and centralized storage, laying the foundation for real-time monitoring by HQ and for the subsequent big data analysis. In contrast to the single machine deployment adopted by Huiyun V1.0 and V2.0, the cloud scheme-based Huiyun V3.0 employs an architecture

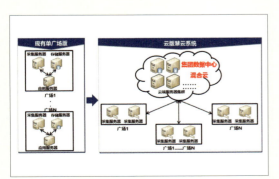

图1 慧云系统3.0版集中运营管控

图2 慧云系统3.0版运营大数据分析

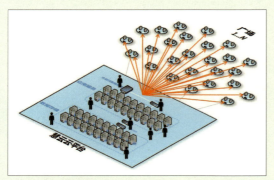

图3 云系统3.0版运营维护简化

with high scalability, high flexibility and centralized O&M. Only this architecture can cater for the demanding requirements of stable operation and long-term mass data storage & analysis, and can support the development and evolution of future system, the system concurrency request can be up to 30 thousand times per second and data storage can be up to 300 billion per year. as shown in Fig. 2. The system concurrency request can be up to 30 thousand times per second and date storage can be up to 300 billion per year.

III. CONVENIENT FOR MAINTENANCE

Huiyun V3.0 excels at unified management and centralized maintenance, so it enables remote and unified upgrade for a number of plazas at HQ cloud, avoiding the version chaos. HQ applies "7× 24" real-time monitoring for cloud platform, so it can promptly provide maintenance once the system goes wrong, and can offer support in the shortest possible time for any problems occurring to the Huiyun system across the country. The system thus enjoys high reliability, as shown in Fig.

IV. WORK PROSPECTS FOR HUIYUN SYSTEM

In the year 2017, Huiyun V3.0 system is in full swing, with all newly opened Wanda plazas being configured with the system. This marks that intelligent management in the Wanda Group Properties for Holding will usher in the cloud era.

In 2017, we will study the specific integration solution of Huiyun, BIM and property management system, to achieve Huiyun V3.0 upgrade, as well as better improve operation management experience and effectiveness. At the same time, we will also study the Huiyun applications in hospitals, small towns, Wanda City and other fields for further improve the overall management quality.

With the consistent perfection of Huiyun System, Wanda Plaza shall have more technical advantages over its competitors in the industry in the aspect of operational management, which will not only help cut down manpower costs, ensure operating quality and reduce operating energy consumption, but also make for Wanda Plaza assets' maintenance and increase of value. With cloud platform-based exploitation on the value of the data and big data analysis, favorable results of further design plan optimization, construction costs reduction, promotion of operational management ability and formation of a virtuous circle among design, construction and operation can be achieved, which shall contribute to the brand promotion of Wanda Plaza and safeguard the smooth implementation of the Group's Asset-light Strategy.

三、"慧云系统"（3.0版）易于简化维护

"慧云"（3.0版）平台具有良好的统一管理及集中维护能力，可在总部"云端"远程、统一对多个广场进行升级，避免了各地广场版本混乱的问题。总部对"云平台"进行"7×24小时"实时监控，一旦系统出现问题，可第一时间进行维护，对于全国各地"慧云"系统运行中发现的问题可在最短时间内在总部端提供支持，确保系统的高可靠性（图3）。

四、"慧云系统"后续工作展望

2017年，万达广场正式全面推广实施"慧云"（3.0版）系统，所有新开业的万达广场均部署"3.0版"的"慧云"系统，万达集团持有物业的智能化管理将开启"云时代"。

2017年，将研究慧云系统与BIM、工程物业管理等系统打通的具体实现方法，实现慧云系统3.0的升级，更好的改善运营管理体验及效果。同时，将研究慧云系统在医院、小镇、万达城等领域的拓展，进一步提升整体的运营管理水平。

"慧云"系统的不断完善，使万达广场在运营管理方面比同行业竞争者拥有更多的技术优势，有助于降低人工成本、保证运行品质、降低运行能耗，有利于万达广场资产的保值、升值。我们将在"云平台"基础上充分挖掘"慧云"系统的数据价值，进行大数据分析，进一步优化设计方案，降低建造成本，提升运营管理能力，设计、建造、运营形成良性循环，有助于万达广场品牌的推广，为集团战略转型的顺利实施保驾护航。

PART G DESIGN & CONTROL 设计及管控

R&D AND OPERATION & MAINTENANCE MANAGEMENT OF HUIYUN SYSTEM V3.0
"慧云系统"（3.0版）的研发及运维管理

| 万达集团总裁助理兼信息管理中心常务副总经理　冯中茜

一、"慧云（3.0）系统"研发工作

2016年，"慧云（3.0）系统"完成了全部产品研发及试点工作，达到了向集团内部提供统一"慧云"产品的目标。"慧云（3.0）"的主要功能的研发及实施开始于2015年中旬，研发过程中同时采用了国际厂商和国内厂商的产品在不同的广场进行研发和试点。集团"慧云工作小组"于2016年4月21日完成对研发成果"入库"评审，经汇报集团"慧云领导小组"，批准并进入后续优化完善及推广试点工作。2016年12月，经过优化完善，并完成推广试点后，"慧云（3.0）"产品正式对外发布，并将应用于所有在2017年后开业的万达广场中。由于包括（3.0）在内的"慧云"各版本对万达集团数字化转型的巨大推动作用，"慧云系统"于2016年12月初，荣获了IDG"2017数字化年度创新全球50强"（2017 Digital Edge 50）大奖，如图1所示。

二、"慧云（3.0）系统"特点

"慧云（3.0）"将系统研发和实施分离，并配套了专门的部署工具以及测试联调工具，在整个部署流程的设计上尽最大可能减少现场实施时对高级技术人员的依赖。现场的系统实施商使用标准化的产品及工具，经过必要的使用培训完成一个广场"慧云系统"的实施工作。

I. R&D OF HUIYUN SYSTEM V3.0

In 2016, the Huiyun System V3.0 has completed the R&D and pilot work for all its products, attaining the goal to provide the Group with a unified Huiyun product. R&D and implementation of Huiyun V3.0 can be dated back to mid-2015, during which products of international manufacturers and domestic manufacturers were used for R&D and test in several plazas. On April 21, 2016, the Group's Huiyun working group completed the List Entry review for the R&D results. After being approved by the Group's Huiyun leading group, the Huiyun System was further optimizing and promoting its pilot projects, which was completed in December 2016. Following that, Huiyun V3.0 was officially released and will be applied to all Wanda Plazas opened after 2017. In early December 2016, Huiyun System has been awarded with the IDG's 2017 Digital Edge 50 for the great role it (Huiyun V3.0 included) plays in driving digital transformation of Wanda Group.

II. CHARACTERISTICS OF HUIYUN SYSTEM V3.0

Huiyun V3.0 boasts of separate R&D and implementation, special deployment tools and joint test debugging tools. The whole deployment flow is designed to minimize dependence on senior technicians when implemented on site. The on-site system implementers use standardized products and tools, and apply Huiyun System to a plaza after necessary operation training.

A "driver library" suitable for equipment of different types, periods and models used in Wanda plazas is set

图1 "慧云系统"广场机房图示

对万达广场所使用的不同类别、不同时期、不同型号的设备建立对接"驱动库",并根据实际情况不断扩充。"驱动库"集成在部署实施工具内,"实施商"对特定设备选择对应的驱动进行后续的实施工作。对于新设备,则由总部研发团队统一研发相关驱动,并扩充"驱动库"。

"慧云(3.0)"的发布及应用,使得万达广场的"慧云系统"实施更加标准化,以统一的数据存储和分析支持公司领导层的决策。一致化的界面方便了广场间的培训及人力资源共享。

三、总部端"混合云"平台

"慧云(3.0)"总部端应用依托建设在万达集团"混合云"平台之上。该平台由软件定义的计算、存储、网络IaaS资源池和PaaS服务组成,并由统一自动化管理系统集中运维提供"云计算"服务。平台底层软件采用业界先进产品,技术成熟可靠,系统全冗余架构,具体多层数据保护和备份机制。"混合云"平台于2016年6月30日验证交付,可支撑2000台虚拟机运行,为后期"慧云"新广场接入扩容奠定良好基础,如图2所示。

"混合云"建设中所包含的PaaS平台是对"慧云(3.0)系统"进行有效支撑的关键。从建设之初,"混合云"PaaS平台就以提供全面完整的服务能力为目标。在统一的服务架构下,建设了Hadoop、HBase、SQLServer、MySQL、消息引擎、检索引擎、认证、主数据、负载均衡等多种能力,所有功能均以服务的方式提供给"慧云(3.0)"使用,极大地降低了系统的研发难度,提升了系统可用性,并且在统一的PaaS平台架构下,各项服务可以后续提供给集团内其他应用系统使用,是万达集团应用系统研发方式的巨大进步。

"慧云(3.0)云版"采用分布式横向水平扩展的基础架构和应用架构,从各广场收集各个子系统数据,然后到总部"混合云"平台中汇总归集,并在大数据平台中进行海量数据的分析和数据建模。"混合云"平台架构完全发挥了"慧云(3.0版)"数据集中及水平

图2 "混合云"平台示意图

up and expanded as the case may be. It is integrated into deployment implementation tools. Implementers selects corresponding equipment driver to carry out the following implementation work. For new equipment, the headquarters R&D team will develop relevant drivers and expand the driver library accordingly.

The release and application of Huiyun V3.0 enable more standardized implementation of Wanda plazas' Huiyun System, support the company's leadership decision-making with unified data storage and analysis, and facilitate training and sharing of human resources among plazas with a uniform interface.

III. HQ HYBRID CLOUD PLATFORM

The Huiyun V3.0 HQ (headquarters) application is based on Wanda Group's "hybrid cloud platform" composed of software-defined computing, storage, network IaaS resource pool and PaaS service. The platform provides cloud computing service through centralized operation & maintenance of centralized automation management system, uses advanced underlying software to have mature and reliable technology, and adopts fully redundant system architecture to ensure multi-tier data protection and backup mechanism. The platform was verified and delivered on June 30, 2016 and it can support 2,000 virtual machines, laying a good foundation for the following expansion with introduction of new Huiyun plazas, as shown in Figure 1.

The PaaS platform included in the "hybrid cloud" is the key to effectively support Huiyun V3.0 System and initially aims to provide comprehensive and complete service capabilities. Under unified service architecture, the platform is configured with diversified functions, including Hadoop, HBase, SQLServer, MySQL, messaging engine, search engine, certification, master data and load balance. All these functions are provided to the Huiyun V3.0 in the form of services, greatly reducing the system's R&D difficulty and improving the system's availability. And thanks to the unified architecture, these services can be shared by remaining application systems of Wanda Group. This marks a huge progress in R&D approach of Wanda's application system.

Huiyun V3.0 adopts an infrastructure and application architecture enabling distributed horizontal scaling. Specifically, it summarizes the subsystem data gathered from plazas at the HQ "hybrid cloud platform" and implements mass data analysis and data modeling at the big data platform. Hybrid cloud platform architecture gives full play to Huiyun V3.0's features of data concentration and horizontally flexible scaling, and better backs healthy running of business and subsystem of Huiyin-based plazas.

IV. O&M SYSTEM AND FLOW OF HUIYUN V3.0 SYSTEM

With view to construction without management is perilous, while developing Huiyun V3.0, Huiyun Working Group and Information Management Center have simultaneously set up O&M management system and flow for the system, so

弹性扩展的特性，更好地支撑"慧云"广场端各个业务及系统子系统的健康运转。

四、"慧云（3.0）系统"运维管理体系及流程

"建而不管则殆"，"慧云（3.0）系统"研发的同时，"慧云工作小组"和信息管理中心就同步建立了系统运维管理体系和流程，以确保信息系统安全、稳定、高效、持续运行。

以制度为先导，2016年12月制订和发布了《"慧云（3.0）系统"运维管理规范》，明确规定各项运维活动的标准流程、《标准操作手册》和岗位职责划分，使运维人员在制度规范和约束下协同操作。《规范》对"慧云"信息系统的故障响应时间、系统可用性、数据备份与恢复、信息安全与防护等方面，提出明确要求。

以工具为手段，利用信息管理中心已有的ITIL流程管理工具，建设了"慧云（3.0）系统"运维管理平台，对各类运维事件全面采集、及时分析和处理、事件工单自动流转、自动记录追踪完成和处理结果。依托现代化运维工具，信息系统的突发事件管理、变更管理、容量管理、发布管理、知识库管理等各项流程管理均实现了标准化和自动化，减少手工重复工作和人为因素干扰。为防止信息系统软硬件故障影响"慧云系统"稳定性，搭建了自动化监控系统，部署业界领先的APM（应用性能监控）产品，实现"慧云（3.0）系统"从广场端到总部全业务"7×24"监控，报警自动通报，故障可以在第一时间被发现和解决，如图3所示。

以团队为保障，完善的管理制度和先进的自动化工具都必须依靠一只高素质的运维团队去落地。"慧云工作小组"挑选和集合总部、各广场以及供应商专业人才，组建了"慧云（3.0）系统"运维团队。通过专项培训、故障模拟演练，逐步提高运维服务团队的专业化水平，有效保障了"慧云系统"的稳定运行。

as to ensure the information system runs in a safe, stable, efficient and continuous manner.

Taking regulation as the precursor: *Specification for Huiyun V3.0 System Operation and Maintenance Management* was formulated and promulgated in December 2016. It clearly stipulates standard flow of O&M activities, *Standard Operation Manual* and division of job responsibilities, enabling O&M personnel to collaboratively operate under institutional norms and constraints. It also sets forth specific requirements on fault response time, system availability, data backup and recovery, information security and protection of Huiyun information system.

Taking tool as a means: O&M management platform of the Huiyun V3.0 System has been built by using the existing ITIL flow management tools in the Information Management Center. The platform enables comprehensive collection, timely analysis and handling, automatic circulation of work orders and automatic recording and tracking of completion and processing results for O&M events. Assisted by modern O&M tools, each flow management, such as emergency management, change management, capacity management, release management and knowledge base management of information system, has been standardized and automated, reducing repetition of manual work and interference of human factors. In order to prevent information system's hardware and software failure from affecting stability of Huiyun System, an automated monitoring system has been set up and configured with the industry-leading APM (Application Performance Monitoring) products. As a result, Huiyun V3.0 System realizes all-business "7 × 24" monitoring of plazas and HQ and automatic notification by alarm, promptly discovering and addressing failures.

Taking teamwork as a guarantee: a high-quality O&M team is indispensable to carry out improved management system and advanced automation tools. Huiyun Working Group has established an O&M team for the Huiyun V3.0 System, with team members being professionals in headquarters, plazas and suppliers. The team will undergo special skill trainings and fault simulation exercises to gradually enhance its professional level, which effectively safeguards stable operation of the Huiyun System.

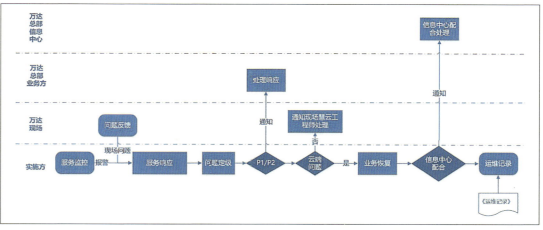

图3 慧云监控系统示意图

BUILD A WORLD-CLASS CULTURAL TOURIST DESTINATION BASED ON REGIONAL CULTURE -INNOVATIVE PRACTICE OF WANDA CULTURAL TOURISM CITY AND WANDA MALL

依托地域文化，打造世界级文化旅游目的地
——万达文化旅游城与万达茂的创新实践

| 万达商业地产总裁助理兼万达商业规划研究院常务副院长　张震

2016年5月南昌万达文化旅游城隆重开业。旅游城"皇冠上的明珠"——南昌万达茂"青花瓷韵"也同时向世人揭开神秘面纱，同年9月。合肥万达文化旅游城也盛大开业——它以极具地域文化特色的形象，如"古籍书卷"一般徐徐展开在徽派粉墙黛瓦之间。

万达文化旅游城和万达茂，凭借其强大的影响力和辐射力，必将对全国文化旅游市场产生震撼影响。

让每个万达文化旅游城成为世界级文化旅游目的地
万达文化旅游城作为世界首创的超大型文化旅游综合体，与以往单一业态主打游乐的主题乐园、度假区都不相同。它不仅有大型的商业中心、休闲酒吧街、豪华五星级酒店，还有大型室外主题乐园、国际大师制作的顶级节目。万达通过自身超强的品牌优势和资源整合能力，将"吃、住、行、游、购、娱"集于一体，推出极具当地文化特色的万达文化旅游城，使游客在其中能感受文化旅游"一站式服务"的全新体验。

多年来，"万达广场就是城市中心"不仅已成为家喻户晓的广告词，更成为"万达广场所在地"人们日常生活中的现实。如今，随着万达文化旅游城遍地开花，人们有理由相信，这些万达文化旅游城，将成为全新的"世界级文化旅游目的地"。

让每个万达茂成为全天候一站式购物体验中心
万达文化旅游城，对中国文化旅游市场无疑是一种创新。万达茂作为万达城中的核心商业综合体，也需要创新。在全国民众日益旺盛的文化旅游需求背景下，如何在商业综合体中引入新的主题娱乐业态、全力打造新产品，成为"万达人"面前的一个全新课题。

万达商业综合体经历了三代的更替。第一代是2001年到2003年的单店模式，第二代是2003到2004年的组合店模式，第三代是2005年以来的大型商业综合体模式。前两代商业模式仍需借助城市传统商圈力量来发展，第三代万达商业综合体已能依靠自身商业模式和规模效应独立创造出城市新中心和新商圈。

在第三代商业综合体的基础上，万达综合研究了迪拜茂、美国茂等成功案例的规划设计模式，在商业综合体中引入各种室内主题娱乐项目，与原有各商业业态互相融合，形成新一代商业综合体、全天候"一站式"

In May 2016, Wanda Cultural Tourism City Nanchang celebrated its official opening, and the Blue-and-White Porcelain Wanda Mall-the crown jewel of tourism city also was unveiled. In September of the same year, Wanda Cultural Tourism City Hefei welcomed its grad opening, unfolding in white walls and black tiles in the shape of ancient scroll, with its distinctive regional cultural features.

Wanda cultural tourism city and Wanda mall are sure to have a presence in the national cultural tourism market, by virtue of their great influence and radial force.

WANDA CULTURAL TOURISM CITY-A WORLD-CLASS CULTURAL TOURIST DESTINATION
As the world's first extra large cultural tourism complex, Wanda cultural tourism city stands out from the traditional amusement-oriented theme parks and resorts focusing on one business type only with its comprehensive inclusion of both large-scale commercial center, leisure bar street, luxury five-star hotel and large outdoor theme park, top programs by the hand of international masters. Benefiting from its own influencing brand advantage and resource integration ability, Wanda introduces Wanda cultural tourism city imbued with local culture characteristics that integrates eating, drinking, amusement, accommodation and shopping into one. Visitors traveling inside may have a fresh one-stop service the cultural tourism brings.

Over the years, "Wanda Plaza is the Heart of the City" is more about a real life in Wanda Plaza location, besides a household advertisement. Nowadays with the blossoming of Wanda cultural tourism city, there are reasons to believe that these cities will emerge as world-class cultural tourist destination.

WANDA MALL-A 24H ONE-STOP SHOPPING EXPERIENCE CENTER
Wanda City is a sure innovation in China's cultural tourism market, so Wanda Mall, being the core commercial complex in Wanda City, also calls for innovation. Against the ever booming cultural tourism demand across the country, our Wanda people are faced with an emerging topic-how to introduce a new theme entertainment business type into commercial complex and how to operate new products.

Wanda commercial complex has had three generations. The first generation is the single-store model from 2001 to 2003; the second generation is the combined-store model from 2003 to 2004; and the third generation is the large commercial-complex model since 2005. The first two models still rely on traditional urban commercial zone, whereas the third model has upgraded to establish new center and new commercial zone of the city depending on its own commercial model and size effect.

购物体验中心——万达茂。

深挖地域文化 成就创新灵感

"继往"才能"开来",只有"有继承的创新",才能行得稳、走得远。万达茂在设计建造过程中,特别注重深挖当地特色历史文化,并将它们在万达项目上焕发新的生命活力,如江西景德镇是中华传统名瓷"青花瓷"的产地,"青花瓷"被人们称为"国家瑰宝"。南昌万达茂巧借这一地域文化,将"青花瓷"形象作为立面设计主体要素(图1),使南昌万达茂成为南昌市乃至江西省的一个文化地标。南昌万达茂"青花瓷"纹饰的创造者、景德镇陶瓷学院何炳钦教授这样评价道:"这样一个长400米、宽200米、高近30米的独特建筑,以26个不同陶瓷形体组合的形态展现在你面前时,你一定会被它的雄伟壮观和独特形态所震撼"。"青花瓷"是古代中华民族智慧的结晶,如今它又成为南昌万达茂设计的灵感来源、核心思想和灵魂所在。

无独有偶,坐落在合肥滨湖新区,南邻巢湖、西望蜀山、依山傍水的合肥万达茂,又如何将当地山光水色以及古徽州的人文气息融入建筑设计中呢?这依然没有难倒独具匠心的"万达人"。以合肥万达茂的主入口为例,设计者把徽州当地书法名家书写的"合肥八景"、"安徽十景",通过图案数字化技术和现代玻璃印刷工艺,在万达茂南北主入口立面上解构重组,将中国文字书法的独特艺术之美与合肥万达茂这一现代建筑完美结合,真正做到了对经典创新演绎,也体现了万达对传统文化的重视与传承(图2)。

立足万达项目 彰显文化自信

万达对文化旅游产业的探索与创新远未停止。随着更多万达城、万达茂的陆续开业,万达将更精准地把握市场脉搏,加快自身创新步伐;通过业态与品牌的有机组合,通过人性化设计和多元化运营,通过更多传统文化内涵的挖掘与运用,使万达城和万达茂不仅成为世界休闲、娱乐、度假和体验式消费的目的地,更成为新时代背景下,展现中国传统文化、让传统文化焕发新生机活力的旗舰产品。

"民族的才是世界的"。万达人相信,这些深具中国特色文化的作品,必将成为中国文化走向世界的"排头兵"。

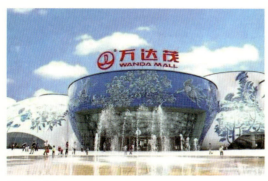

图1 南昌万达茂

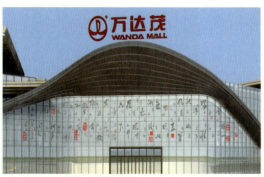

图2 合肥万达茂

Proceeding from the third generation and studying the planning & design model unique to Dubai Mall, U.S.A Mall and other successful cases, Wanda has introduced various indoor theme entertainments into commercial complex, which works in harmony with the original business types. Accordingly, Wanda Mall - A 24H One-Stop Shopping Experience Center, a new generation commercial complex takes its shape.

EXPLORE REGIONAL CULTURE, GAIN INNOVATIVE INSPIRATION

Conting with the past is to open up in the future; only innovate with inheritance can we go steadily and go far. When designing and constructing Wanda mall, we have delved into local featured history and culture and made them invigorated in Wanda projects. Considering that Jingdezhen in Jiangxi Province is the origin of China's traditional Blue-and-White Porcelain, the national treasure. Nanchang Wanda Mall utilized this regional culture and largely applied blue and white porcelain image in its facade design, making itself a cultural landmark in Nanchang and even the Jiangxi Province. Mr. He Bingqin, a professor at Jingdezhen Ceramic Institute and the creator of Nanchang Wanda Mall's Blue-and-White Porcelain pattern commented: "Seeing such a 400m×200m×30m architecture with combination of 26 ceramic shapes, one is surely be impressed by its spectacular and unique form". Blue and white porcelain gathering ancient Chinese wisdom has now emerged as inspiration source, core idea and soul of Nanchang Wanda Mall's design.

Similarly, the scenic Hefei Wanda Mall is located in Binhu New District, Hefei, facing Chaohu Lake to the south and Mountain Shushan to the west. How to incorporate the local scenery and ancient Huizhou's culture into architectural design? Wanda Mall Hefei's main entrance, for example, has deconstructed and restructured Huizhou calligraphy masters' writing of "Hefei Eight Sights" and "Anhui Ten Sights" on facades of the south and north main entrances, by means of pattern digitalization technology and modern glass printing process. Such design perfectly blends unique art beauty of Chinese calligraphy with modern building, truly innovating the classics and demonstrating Wanda's emphasis and inheritance on traditional culture.

DO WANDA PROJECTS, DEMONSTRATE CULTURAL CONFIDENCE

Wanda will never stop its exploration and innovation in its cultural tourism industry. With several more Wanda cities and Wanda malls being opened, Wanda will exactly catch up with the current trend of the market and accelerate its own pace of innovation. By virtue of organic combination between business types and brands, humanized design and diversified operation, as well as deep exploration and application of traditional culture connotation, Wanda cities and Wanda malls will grow to be the world's destination for leisure, entertainment, holiday and experiential consumption, and moreover, a flagship product that shows and invigorates the Chinese traditional culture against the new era.

As the saying goes that "What Ethnic is What Worldwide", Wanda People believe, these works imbued with Chinese culture characteristics will serve as vanguards to lead Chinese culture to go global.

WANDA IN THE EYES OF ALL CIRCLES
各界看万达

"概念商业广场"国际建筑设计竞赛颁奖典礼落幕

中国建筑文化研究会副会长、秘书长刘凌宏先生在颁奖典礼致辞中说道:"2016是'十三五'开局之年,'十三五'提出了'创新、协调、绿色、开放、共享'的新发展理念。创新是引导发展的第一动力,城市的发展更需要创新。在快速发展的互联网时代,使人们走出孤独的是交流体验;虚拟与实体之间不再是对抗,而演变成一种共生。螺旋上升的轨迹一同共进,抵达的彼岸应该是以人为本的新体验"。刘凌宏先生强调,"这次竞赛的参赛作品,确实有很多想法和概念已经颠覆了我们对一些商业广场现有的固有思维模式,很多都非常优秀。万达集团一直在不断地以创新探索的精神走出了万达自己的模式。因此,这样的国际竞赛提供了一个面向全球的开放平台,凸现出'万达智造'的优势,集广大优秀者的智慧,以商业广场为探讨的起点,让更多创新意识最大地发挥社会价值,更好的理念在此凝聚为更大的力量,共同探索出适应中国城市发展的道路。"

(新华网,2016年05月04日)

INTERNATIONAL SHOPPING PLAZA CONCEPT COMPETITION AWARDING CEREMONY CONCLUDED

In his speech at the award ceremony, Mr. Liu Linghong, Vice Chairman and Secretary-General of Architecture and Culture Society of China (ACSC) said, "2016 is the beginning year of the '13th Five-year Plan', which comes up with the new development idea of 'innovation, coordination, ecology, openness and sharing'. Innovation is the first impetus that drives development, and is of special significance in the development of a city. In the fast growing times of Internet, Communication and experiences gets people out of loneliness. Virtuality is no more opposite to the real world; instead harmonious intergrowth is seen between them. Moving forward together in a spiral trajectory, they will create people-oriented new experiences." Liu stressed that "a lot of excellent works are presented in the contest, bringing us ideas and concepts that overturned our thoughts about shopping malls. Wanda Group has been keeping innovation and exploration to establish its unique mode of development. Such an international competition provides us an open global platform to show our advantages 'made with Wanda's wisdom', and, by the intelligence of the mass talented people and starting from shopping malls, to explore how to maximize the social values of more sense of innovation and how to transform better ideas into greater strengths, so as to jointly carve a development path suitable for Chinese cities."

(http://xinhuanet.com, May 4th, 2016)

合肥市委书记吴存荣在合肥万达城开业讲话摘录

"万达文化旅游城是兼具世界眼光和中国特色的精品,也是融合地方元素和高科技的精品作品,是现代气派和年轻潮流的呈现,必然会引领行业发展的新标杆、彰显合肥特色品质的新地标、服务民生社会的新窗口"。

(2016年09月24日)

EXCERPT OF SPEECH BY WU CUNRONG, HEFEI MUNICIPAL PARTY COMMITTEE SECRETARY AT THE OPENING CEREMONY OF WANDA CITY HEFEI

"Wanda Cultural Tourism City is an elaborated product designed with both international vision and Chinese characteristics. It also integrates local elements and high technologies, as well as modern qualities and fresh trends, and will inevitably become a new development standard for the industry, a new landmark that highlights distinctive qualities of Hefei, and a new window that serves the people's well-being and the society."

(September 24th, 2016)

万达城复制提速,加快业务转型

合肥万达文化旅游城包括"水乐园、电影乐园和主题乐园"三大板块。其中,室内水乐园建筑面积达3.2万平方米,是目前世界规模和技术领先的第四代室内恒温主题水乐园之一。合肥万达文化旅游城还包括安徽规模空前的星级酒店群,5家高等级酒店环绕着18万平方米的景观湖而建。其中万达文华酒店是万达自主创立并管理的中国第一座城市度假酒店。

进军文化旅游是万达业务转型的初衷。由于迪士尼在国内开业火爆,万达的文化旅游产品频频被拿来与迪士尼进行比较,话题不断。王健林曾对外表示,上海只有一个迪士尼,万达在全国其他地方,可开15到20个。"我们觉得安徽地区特别适合发展旅游业,当地也很支持。合肥万达文化旅游城现在开业,大部分项目都是我们自主创新研发的,而且还融合了当地徽派文化,体现在演艺节目和建筑中"——王健林在开业仪式现场如是说。

(第一财经日报,2016年09月27日)

ACCELERATION OF WANDA CITY'S DUPLICATION AIMING AT QUICKER BUSINESS TRANSFORMATION

Wanda Cultural Tourism City Hefei consists of three major parts, namely the water park, movie park and theme park. Among them, the indoor water park covers a floor area of 32,000 square meter, one of the world's largest and fourth-generation indoor heated theme water parks with most advanced technologies. Wanda Cultural Tourism City Hefei also has a star hotel complex with an unprecedented scale in Anhui Province, where 5 high-grade hotels are distributed surrounding a 180,000 square meter scenic lake. Among those hotels, Vista Wanda is the first urban resort hotel that was founded and is operated all by Wanda.

The original purpose of Wanda business transformation is aimed at cultural and tourism industy. Its cultural tourism products have been repeatedly compared with those of Disneyland, and has been a hot topic of debate. According to Wang Jianlin, there is only one Disneyland park in Shanghai, but up to 15 to 20 Wanda cities might be opened up in the other parts of China. "We think that Anhui is particularly suitable for tourism development, and have got a lot of local supports. Most programs of the cultural tourism city opened just now were developed by us independently, with local Hui-style cultural elements incorporated in the performing arts programs and architecture", said Wang Jianlin at the opening ceremony.

(First Financial Daily, September 27th, 2016)

王健林的"梦想之城":依托中国文化魅力"斗法"迪士尼

2016年9月24日,合肥万达文化旅游城正式开业运营。这是年内开业的第二座"万达城"。4个月前,首个以"万达城"命名的超大型文旅商综合项目"南昌万达城"正式开业。截至目前,万达开业的大型文化旅游项目共有五个。尽管开业当天天气稍显炎热,然而合肥万达城还是吸引了从四面八方赶来的游客,超级乐园一开业就上演"短兵相接",晚上十点的万达茂依旧人头攒动。

从万达城这支"温度计"背后,我们可以感知到中国的文化潜力、旅游和消费需求的真实体温。"合肥是目前旅游投资最好的城市之一",王健林说,"目前万达非常看好合肥的未来发展前景。""合肥是国家级的高铁枢纽,从全国各地到合肥都非常便捷,这个城市在全国的地位和影响力都正在迅速提升。"

记者注意到,合肥万达城还有一个鲜明的特点,就是围绕"水"做了不少文章。无论运用科技手段的"巢湖探秘",还是喷泉与水上表演的完美演绎,抑或者是建筑面积达3.2万平方米的室内水乐园,都极具震撼效果,这与合肥打造"大湖名城、创新高地"规划高度契合。

随着新一轮消费升级,旅游如今已经成为国民日常生活的"刚需"。统计显示,2015年,中国国内旅游突破40亿人次,旅游收入过4万亿元人民币;出境旅游1.2亿人次。中国国内旅游、出境旅游人次和国内旅游消费、境外旅游消费均列世界第一。

合肥万达城的开业,将力争使得安徽成为中部区域旅游的中心,激活当地和周边的旅游需求。万达城同时还是一个涵盖多个板块的多元业态,除了"主题乐园、水乐园、电影乐园"之外,周边环绕的万达茂、度假酒店群、酒吧街、特色餐馆等都将旅游的产业链伸展拉长。

(新华网,2016年09月27日)

WANG JIANLIN'S "DREAM CITY": FIGHTING AGAINST DISNEYLAND WITH CHINESE CULTURE

On September 24th, 2016, Wanda Cultural Tourism City Hefei opened to the public officially. It is the second "Wanda City" opened in the year. 4 months ago, Wanda City Nanchang, the first super-scale-tourism-commercial complex first named as "Wanda City" was officially opened. As of now, altogether five large-scale cultural tourism projects have been in operation. Despite of the slightly hot weather on the opening day, tourists came from all sides to Wanda City Hefei. The super park attracted high attention on the opening day and remained very lively until late on the night.

Wanda City is just like a "thermometer" that reflects the cultural potential, tourism and consumer demands in China. "Hefei is one of the best cities for tourism investment in China today", Wang Jianlin said, "We are optimistic about its development opportunities". "It is a national high-speed rail hub, with convenient transportation to anywhere else in the country; it has fast-improving status and influences in China."

We noticed that "water" is one of the distinct characteristics of Wanda City Hefei and frequently utilized in its design. Whether in "Exploration in Chaohu Lake" featuring technological means, the perfect fountains and water performance, or the indoor water park with a floor area up to 32,000 square meter, people will be shocked by the magnificent effects, which exactly matches the goal of Hefei's planning to "Build Great Lake City and Innovative Highland".

With the new round of consumption upgrade, tourism has become the "rigid demand" in the daily life of Chinese people. Statistics show that in 2015, the domestic tourism in China exceeded 400 million, with an annual income over 4 trillion Yuan; outbound tourism exceeded 120 million. China ranked first in the world in domestic and outbound tourism, as well as domestic tourism consumption and outbound tourism consumption.

Wanda City Hefei will catalyze Anhui to become the tourism center of central China, and to stimulate tourist demands in both local and surrounding areas. Besides the above, Wanda City also includes various business types. In addition to "Theme Park, Water Park and Movie Park", there is also Wanda Mall, resort hotel complex, bar street, featured restaurants, etc., expanding the tourism industry chain.

(http://xinhuanet.com, September 27th, 2016)

万达"双城"首迎黄金周,爆满客流印证王健林底气

当前国内旅游出行市场仍是一片大蓝海。百姓日益增长的休闲、度假等多元需求与高质量、多业态的供给呈现强烈的矛盾。即便在"北上广深"这样的大都市,能够满足百姓休闲、娱乐、购物、度假等丰富需求的大型休闲旅游目的地也是凤毛麟角,更别说是诸多二三线城市。

从已经开业的南昌万达城、合肥万达城来看,业态丰富度上远超传统意义上的度假区,包括大型室外乐园、超大型的Mall、结合城市特点配置的室内乐园(譬如水上乐园、海洋馆等)、世界级的酒店群、酒吧街,以丰富的业态满足了国人多层次的休闲度假需求,更为关键的是其可复制性与高速扩张性。

在努力解决"有"的同时,王健林也在试图满足"优"这个难题,即构建属于中国的文化竞争力。王健林深谙,文化产品仅靠简单的"复制、粘贴"是行不通的。因此,从万达城的设计不难看出,"就地取材、本土特色和当地文化"让每座万达城都别具一格。他希望,这些伴随国人成长的故事和元素能唤起消费者的共鸣,从而带给大家快乐。

(人民网,2016年10月11日)

WANDA "TWIN CITIES" WELCOMING THE FIRST GOLDEN WEEK, CROWDED CONSUMERS SHOW WHY WANG JIANLIN IS SO CONFIDENT

Currently, domestic travel market is still a vast "blue ocean". There are strong contradictions between people's growing diversified demands for recreation, vacationing, etc. and the high-quality, multi-format supplies. Even in large cities like Beijing, Shanghai, Guangzhou and Shenzhen, only a few large-scale recreational and tourism destinations are available to meet people's rich needs for recreation, entertainment, shopping, vacationing and so on, not to mention the numerous second- or third-tier cities.

Take Wanda City Nanchang and Wanda City Hefei for example, they provide far more business types than traditional resorts. They have large-sized outdoor park, super large mall, indoor park with city characteristics (water park, aquarium, etc.), world-class hotel complex, bar street and so on. Thus, the people's recreation and vacationing demands could be met in a way that is highly replicable and fast-expandable.

While solving the problem of changing 0 to 1, Wang Jianlin is also try to make every project excellent, or in other words, to build up the unique cultural competitiveness of China. He knows very well it's not enough to promote cultural products by simply "copying and pasting". As a result, the design of each Wanda City reflects distinct "local materials, local characteristics and local cultures" to make it unique. He hopes these stories and elements that have accompanied the Chinese people's whole life will strike a chord inside them and bring them happiness.

(People's Daily, October 11th, 2016)

项目索引 PROJECT INDEX

INDEX OF WANDA CITYS
万达城索引

01 WANDA MALL NANCAHNG
南昌万达茂

大商业施工图设计单位	中国电子工程设计院
外立面设计单位	上海鼎实建筑设计有限公司
内装设计单位	北京清尚建筑设计研究院有限公司
景观设计单位	华东建筑设计研究院有限公司
导向标识设计单位	北京视域四维城市导向系统规划设计有限公司
夜景照明设计单位	深圳市千百辉照明工程有限公司
弱电智能化设计单位	上海中电电子系统工程有限公司
外幕墙深化设计单位	厦门开联装饰工程有限公司

02 WANDA VISTA RESORT NANCHANG
南昌万达文华度假酒店

大商业施工图设计单位	上海同济设计研究院
外立面设计单位	施坦伯格建筑咨询（上海）有限公司
内装设计单位	万达酒店设计研究院
景观设计单位	上海帕莱登建筑景观咨询有限公司
导向标识设计单位	北京视域四维城市导向系统规划设计有限公司
夜景照明设计单位	深圳市金达照明股份有限公司
弱电智能化设计单位	万达酒店设计研究院
外幕墙深化设计单位	上海旭密林幕墙有限公司

03 WANDA REALM RESORT NANCHANG
南昌万达嘉华度假酒店

大商业施工图设计单位	上海同济设计研究院
外立面设计单位	何弢设计国际（上海）有限公司
内装设计单位	万达酒店设计研究院
景观设计单位	上海帕莱登建筑景观咨询有限公司
导向标识设计单位	北京视域四维城市导向系统规划设计有限公司
夜景照明设计单位	深圳市金达照明股份有限公司
弱电智能化设计单位	万达酒店设计研究院
外幕墙深化设计单位	上海旭密林幕墙有限公司

04 PULLMAN NANCHANG WANDA
南昌万达铂尔曼酒店

大商业施工图设计单位	上海同济设计研究院
外立面设计单位	何弢设计国际（上海）有限公司
内装设计单位	万达酒店设计研究院
景观设计单位	上海帕莱登建筑景观咨询有限公司
导向标识设计单位	北京视域四维城市导向系统规划设计有限公司
夜景照明设计单位	深圳市金达照明股份有限公司
弱电智能化设计单位	万达酒店设计研究院
外幕墙深化设计单位	上海旭密林幕墙有限公司

05 NOVOTEL NANCHANG WANDA
南昌万达诺富特酒店

大商业施工图设计单位	上海中建设计研究院
外立面设计单位	施坦伯格建筑咨询（上海）有限公司
内装设计单位	万达酒店设计研究院
景观设计单位	上海帕莱登建筑景观咨询有限公司
导向标识设计单位	北京视域四维城市导向系统规划设计有限公司
夜景照明设计单位	深圳市金达照明股份有限公司
弱电智能化设计单位	万达酒店设计研究院
外幕墙深化设计单位	北京金星卓宏幕墙有限公司

06 MERCURE NANCHANG WANDA
南昌万达美居酒店

大商业施工图设计单位	上海中建设计研究院
外立面设计单位	施坦伯格建筑咨询（上海）有限公司
内装设计单位	万达酒店设计研究院
景观设计单位	上海帕莱登建筑景观咨询有限公司
导向标识设计单位	北京视域四维城市导向系统规划设计有限公司
夜景照明设计单位	深圳市金达照明股份有限公司
弱电智能化设计单位	万达酒店设计研究院
外幕墙深化设计单位	北京金星卓宏幕墙有限公司

07 WANDA MALL HEFEI
合肥万达茂

大商业施工图设计单位	华东建筑设计研究院有限公司
外立面设计单位	上海新外建工程设计与顾问有限公司
内装设计单位	北京清尚建筑设计研究院有限公司
景观设计单位	北京中建建筑设计院上海分院
导向标识设计单位	北京清尚建筑设计研究院有限公司
夜景照明设计单位	深圳市千百辉照明工程有限公司
弱电智能化设计单位	上海智信世创智能集成有限公司
外幕墙深化设计单位	上海旭密林幕墙有限公司

08 WANDA VISTA HEFEI
合肥万达文华酒店

大商业施工图设计单位	中国建筑上海设计研究院有限公司
外立面设计单位	贝加艾奇（上海）建筑设计咨询有限公司（B+H）
内装设计单位	万达酒店设计研究院
景观设计单位	IMG3（上海）矶森景观规划设计事务所
导向标识设计单位	北京艺同博雅企业形象设计有限公司
夜景照明设计单位	中国建筑装饰集团有限公司
弱电智能化设计单位	万达酒店设计研究院
外幕墙深化设计单位	上海旭密林幕墙有限公司

09 WANDA REALM HEFEI
合肥万达嘉华酒店

大商业施工图设计单位	同济大学建筑设计研究院（集团）有限公司
外立面设计单位	施坦伯格建筑咨询（上海）有限公司
内装设计单位	万达酒店设计研究院
景观设计单位	IMG3（上海）矶森景观规划设计事务所
导向标识设计单位	北京艺同博雅企业形象设计有限公司
夜景照明设计单位	中国建筑装饰集团有限公司
弱电智能化设计单位	万达酒店设计研究院
外幕墙深化设计单位	北京嘉寓门窗幕墙股份有限公司

10 PULLMAN HEFEI WANDA
合肥万达铂尔曼酒店

大商业施工图设计单位	同济大学建筑设计研究院（集团）有限公司
外立面设计单位	上海日清建筑设计有限公司
内装设计单位	万达酒店设计研究院
景观设计单位	IMG3（上海）矶森景观规划设计事务所
导向标识设计单位	北京艺同博雅企业形象设计有限公司
夜景照明设计单位	中国建筑装饰集团有限公司
弱电智能化设计单位	万达酒店设计研究院
外幕墙深化设计单位	北京嘉寓门窗幕墙股份有限公司

11 NOVOTEL HEFEI WANDA
合肥万达诺富特酒店

大商业施工图设计单位	中国建筑上海设计研究院有限公司
外立面设计单位	北京欧安地合众建筑设计顾问有限公司（OAD）
内装设计单位	万达酒店设计研究院
景观设计单位	IMG3（上海）矶森景观规划设计事务所
导向标识设计单位	北京艺同博雅企业形象设计有限公司
夜景照明设计单位	中国建筑装饰集团有限公司
弱电智能化设计单位	万达酒店设计研究院
外幕墙深化设计单位	上海旭密林幕墙有限公司

12 MERCURE HEFEI WANDA
合肥万达美居酒店

大商业施工图设计单位	中国建筑上海设计研究院有限公司
外立面设计单位	施坦伯格建筑咨询（上海）有限公司
内装设计单位	万达酒店设计研究院
景观设计单位	IMG3（上海）矶森景观规划设计事务所
导向标识设计单位	北京艺同博雅企业形象设计有限公司
夜景照明设计单位	中国建筑装饰集团有限公司
弱电智能化设计单位	万达酒店设计研究院
外幕墙深化设计单位	上海旭密林幕墙有限公司

INDEX OF WANDA PLAZAS
万达广场索引

01 CHENGDU SHUDU WANDA PLAZA
成都蜀都万达广场

大商业施工图设计单位	中国建筑西南设计研究院有限公司
外立面设计单位	艾奕康设计与咨询（深圳）有限公司上海分公司
内装设计单位	北京清尚建筑设计研究院有限公司
景观设计单位	上海中星志成建筑设计有限公司
导向标识设计单位	北京艺同博雅企业形象设计有限公司
夜景照明设计单位	北京鱼禾光环境设计有限公司
弱电智能化设计单位	大连理工科技有限公司
外幕墙深化设计单位	中国建筑科学研究院

02 HUBEI JINGMEN WANDA PLAZA
湖北荆门万达广场

大商业施工图设计单位	北京炎黄联合国际工程设计有限公司
外立面设计单位	上海联创建筑有限公司
内装设计单位	中国中建设计集团有限公司
景观设计单位	福建泛亚远景环境设计工程有限公司
导向标识设计单位	北京视域四维城市导向系统规划设计有限公司
夜景照明设计单位	栋梁国际照明（北京）中心有限公司
弱电智能化设计单位	华体集团有限公司
外幕墙深化设计单位	北京和平幕墙工程有限公司

03 HUNAN XIANGTAN WANDA PLAZA
湖南湘潭万达广场

大商业施工图设计单位	中国中轻国际工程有限公司
外立面设计单位	华凯帕特建筑设计（北京）有限公司
内装设计单位	北京筑邦建筑装饰工程有限公司
景观设计单位	重庆浩丰规划设计集团股份有限公司
导向标识设计单位	北京艺同博雅企业形象设计有限公司
夜景照明设计单位	北京三色石环境艺术设计有限公司
弱电智能化设计单位	北京益泰牡丹电子工程有限责任公司
外幕墙深化设计单位	深圳海外装饰工程有限公司

04 TAIZHOU JINGKAI WANDA PLAZA
台州经开万达广场

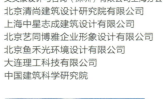

大商业施工图设计单位	华汇工程设计集团股份有限公司
外立面设计单位	华汇工程设计集团股份有限公司
内装设计单位	上海浦东建筑设计研究院有限公司
景观设计单位	上海中星志成建筑设计有限公司
导向标识设计单位	北京视域四维城市导向系统规划设计有限公司
夜景照明设计单位	北京鱼禾光环境设计有限公司
弱电智能化设计单位	上海智信世创智能系统集成有限公司
外幕墙深化设计单位	华汇工程设计集团股份有限公司

05 ZIYANG WANDA PLAZA
资阳万达广场

大商业施工图设计单位	北京炎黄联合国际工程设计有限公司
外立面设计单位	沪张思建筑设计咨询（上海）有限公司
内装设计单位	中国中建设计集团有限公司
景观设计单位	深圳市铁汉生态环境股份有限公司
导向标识设计单位	北京视域四维城市导向系统规划设计有限公司
夜景照明设计单位	深圳市金达照明有限公司
弱电智能化设计单位	北京亚洲卫星通信技术有限公司
外幕墙深化设计单位	北京国科天创建筑设计院有限责任公司

06 JI'NAN HIGH-TECH WANDA PLAZA
济南高新万达广场

大商业施工图设计单位	山东同圆设计集团有限公司
外立面设计单位	中国新兴建设开发总公司
内装设计单位	深圳三九装饰工程有限公司
景观设计单位	华汇工程设计集团股份有限公司
导向标识设计单位	北京艺同博雅企业形象设计有限公司
夜景照明设计单位	北京三色石环境艺术设计有限公司
弱电智能化设计单位	上海智信世创智能系统集成有限公司
外幕墙深化设计单位	深圳市华辉装饰工程有限公司

07 WUHAI WANDA PLAZA
乌海万达广场

大商业施工图设计单位	中国中轻国际工程有限公司
外立面设计单位	亨派建筑设计咨询（上海）有限公司
内装设计单位	北京市建筑装饰设计院有限公司
景观设计单位	福建泛亚远景环境设计工程有限公司
导向标识设计单位	北京视域四维城市导向系统规划设计有限公司
夜景照明设计单位	上海易照景观设计有限公司
弱电智能化设计单位	南京熊猫信息产业有限公司
外幕墙深化设计单位	北京市金星卓宏幕墙工程有限公司

08 GUANGDONG ZHANJIANG WANDA PLAZA
广东湛江万达广场

大商业施工图设计单位	奥意建筑工程设计有限公司
外立面设计单位	北京五合国际建筑设计咨询有限公司
内装设计单位	深圳市三九装饰工程有限公司
景观设计单位	深圳市致道景观有限公司
导向标识设计单位	北京艺同博雅企业形象设计有限公司
夜景照明设计单位	深圳普莱思照明设计顾问有限责任公司
弱电智能化设计单位	北京国安电气有限责任公司
外幕墙深化设计单位	北京市金星卓宏幕墙工程有限公司

09 SIPING WANDA PLAZA
四平万达广场

大商业施工图设计单位	大连市建筑设计研究院有限公司
外立面设计单位	豪斯泰勒张思图德建筑设计咨询（上海）有限公司
内装设计单位	北京清尚建筑设计研究院有限公司
景观设计单位	深圳文科园林股份有限公司
导向标识设计单位	北京艺同博雅企业形象设计有限公司
夜景照明设计单位	北京市洛西特灯光设计顾问有限公司
弱电智能化设计单位	大连理工科技有限公司
外幕墙深化设计单位	深圳市新山幕墙技术咨询有限公司

10 JIXI WANDA PLAZA
鸡西万达广场

大商业施工图设计单位	哈尔滨工业大学建筑设计研究院
外立面设计单位	亚瑞建筑设计有限公司
内装设计单位	北京城建长城建筑装饰工程有限公司
景观设计单位	福建泛亚远景环境设计工程有限公司
导向标识设计单位	北京广音德视觉技术股份有限公司
夜景照明设计单位	深圳市金达照明有限公司
弱电智能化设计单位	南京熊猫信息产业有限公司
外幕墙深化设计单位	北京市金星卓宏幕墙工程有限公司

11 MUDANJIANG WANDA PLAZA
牡丹江万达广场

大商业施工图设计单位	哈尔滨工业大学建筑设计研究院
外立面设计单位	北京中外建建筑设计有限公司
内装设计单位	北京筑邦建筑装饰工程有限公司
景观设计单位	中国建筑设计院有限公司
导向标识设计单位	北京艺同博雅企业形象设计有限公司
夜景照明设计单位	深圳市金达照明有限公司
弱电智能化设计单位	华体集团有限公司
外幕墙深化设计单位	北京国科天创建筑设计院有限责任公司

12 TONGLIAO WANDA PLAZA
通辽万达广场

大商业施工图设计单位	北京东方国兴建筑设计有限公司
外立面设计单位	北京东方国兴建筑设计有限公司
内装设计单位	上海浦东建筑设计研究院有限公司
景观设计单位	上海兴田建筑工程设计事务所
导向标识设计单位	北京艺同博雅企业形象设计有限公司
夜景照明设计单位	北京市建筑设计研究院有限公司
弱电智能化设计单位	北京益泰牡丹电子工程有限责任公司
外幕墙深化设计单位	北京和平幕墙工程有限公司

13 XINING WANDA PLAZA 西宁万达广场

大商业施工图设计单位	中国电子工程设计院有限公司
外立面设计单位	上海鼎实建筑设计有限公司
内装设计单位	广州集美设计工程有限公司
景观设计单位	上海兴田建筑工程设计事务所
导向标识设计单位	北京广育德视觉技术有限公司
夜景照明设计单位	北京三色石环境艺术设计院有限公司
弱电智能化设计单位	大连理工科技有限公司
外幕墙深化设计单位	北京金星卓宏幕墙工程有限公司

14 SHIYAN WANDA PLAZA 十堰万达广场

大商业施工图设计单位	北京柯德普建筑设计顾问有限公司
外立面设计单位	北京柯德普建筑设计顾问有限公司
内装设计单位	中国中建设计集团有限公司
景观设计单位	华东建筑设计研究院有限公司
导向标识设计单位	北京广育德视觉技术有限公司
夜景照明设计单位	深圳市金达照明有限公司
弱电智能化设计单位	北京国安电气有限责任公司
外幕墙深化设计单位	湖北弘毅建设有限公司

15 BOZHOU WANDA PLAZA 亳州万达广场

大商业施工图设计单位	安徽省建筑设计研究院有限公司
外立面设计单位	北京欧安地合众设计顾问有限公司
内装设计单位	深圳市三九装饰工程有限公司
景观设计单位	北京中建设计院有限公司上海分公司
导向标识设计单位	北京艺同博雅企业形象设计有限公司
夜景照明设计单位	央美光成（北京）建筑设计有限公司
弱电智能化设计单位	北京益泰牡丹电子工程有限公司
外幕墙深化设计单位	中国新兴建设开发总公司

16 URUMQI JINGKAI WANDA PLAZA 乌鲁木齐经开万达广场

大商业施工图设计单位	中信建筑设计院有限公司
外立面设计单位	卡斯帕建筑设计咨询（上海）有限公司
内装设计单位	中国中建设计集团有限公司
景观设计单位	深圳文科园林股份有限公司
导向标识设计单位	北京四维城市导向系统规划设计有限公司
夜景照明设计单位	北京鱼禾光环境设计有限公司
弱电智能化设计单位	北京益泰牡丹电子工程有限责任公司
外幕墙深化设计单位	北京和平幕墙工程有限公司

17 HEFEI YAOHAI WANDA PLAZA 合肥瑶海万达广场

大商业施工图设计单位	中国电子工程设计院有限公司
外立面设计单位	上海鼎实建筑设计有限公司
内装设计单位	广东省集美设计工程有限公司
景观设计单位	中国城市建设研究院有限公司
导向标识设计单位	深圳市上行线设计有限公司
夜景照明设计单位	艾ескоa联合照明设计（北京）有限公司
弱电智能化设计单位	北京亚洲卫星通信技术有限公司
外幕墙深化设计单位	北京国科天创建筑设计院有限责任公司

18 YANJI WANDA PLAZA 延吉万达广场

大商业施工图设计单位	大连都市发展设计有限公司
外立面设计单位	华凯建筑设计（上海）有限公司
内装设计单位	北京市建筑装饰设计院有限公司
景观设计单位	鲁迅美术学院艺术工程总公司
导向标识设计单位	北京广育德视觉技术有限公司
夜景照明设计单位	北京东方国兴照明有限公司
弱电智能化设计单位	大连理工科技有限公司
外幕墙深化设计单位	北京和平幕墙工程有限公司

19 PANJIN WANDA PLAZA 盘锦万达广场

大商业施工图设计单位	北京柯德普建筑设计顾问有限公司
外立面设计单位	深圳市新山幕墙技术咨询有限公司
内装设计单位	上海浦东建筑设计研究院有限公司
景观设计单位	中景汇景观设计有限公司
导向标识设计单位	北京罗丹莫纳牌业有限公司
夜景照明设计单位	北京鱼禾光环境设计有限公司
弱电智能化设计单位	北京益泰牡丹电子工程有限责任公司
外幕墙深化设计单位	北京建黎铝门窗幕墙有限公司

20 YIWU WANDA PLAZA 义乌万达广场

大商业施工图设计单位	中国中轻国际工程有限公司
外立面设计单位	华凯建筑设计（上海）有限公司
内装设计单位	北京清尚建筑研究院有限公司
景观设计单位	华东建筑设计研究院有限公司
导向标识设计单位	北京艺同博雅企业形象设计有限公司
夜景照明设计单位	深圳市标美照明设计有限公司
弱电智能化设计单位	上海延华智能科技（集团）股份有限公司
外幕墙深化设计单位	深圳市新山幕墙技术咨询有限公司

21 CHONGQING YONGCHUAN WANDA PLAZA 重庆永川万达广场

大商业施工图设计单位	中国建筑西南设计研究院有限公司
外立面设计单位	北京华雍汉维建筑咨询有限公司
内装设计单位	中国中建设计集团有限公司
景观设计单位	华汇工程设计集团股份有限公司
导向标识设计单位	北京视域四维城市导向系统规划设计有限公司
夜景照明设计单位	深圳市千百辉照明有限公司
弱电智能化设计单位	上海智信世创智能系统集成有限公司
外幕墙深化设计单位	厦门嘉福幕墙铝窗有限公司

22 CHANGDE WANDA PLAZA 常德万达广场

大商业施工图设计单位	北京国科天创建筑设计院有限责任公司
外立面设计单位	北京华雍汉维建筑咨询有限公司
内装设计单位	北京市建筑装饰设计院有限公司
景观设计单位	北京中建建筑设计院有限公司上海分公司
导向标识设计单位	大连依斯特图文导视设计工程有限公司
夜景照明设计单位	北京鱼禾光环境设计有限公司
弱电智能化设计单位	北京国安电气有限责任公司
外幕墙深化设计单位	湖北弘毅建设有限公司

23 DEYANG WANDA PLAZA 德阳万达广场

大商业施工图设计单位	成都基准方中建筑设计有限公司
外立面设计单位	艾奕康建筑设计（深圳）有限公司上海分公司
内装设计单位	广东省集美设计工程有限公司
景观设计单位	上海帕莱登建筑景观咨询有限公司
导向标识设计单位	深圳市上行线设计有限公司
夜景照明设计单位	北京三色石环境艺术设计院有限公司
弱电智能化设计单位	浙江远望科技有限公司
外幕墙深化设计单位	北京市金星卓宏幕墙工程有限公司

24 SICHUAN LESHAN WANDA PLAZA 四川乐山万达广场

大商业施工图设计单位	北京柯德普建筑设计顾问有限公司
外立面设计单位	北京国科天创建筑设计院有限责任公司
内装设计单位	北京市建筑装饰设计院有限公司
景观设计单位	北京中建设计院有限公司上海分公司
导向标识设计单位	北京视域四维城市导向系统规划设计有限公司
夜景照明设计单位	艾德联合照明设计（北京）有限公司
弱电智能化设计单位	南京熊猫信息产业有限公司
外幕墙深化设计单位	北京国科天创建筑设计院有限责任公司

25 MEIZHOU WANDA PLAZA 梅州万达广场

大商业施工图设计单位	深圳市同济人建筑设计有限公司
外立面设计单位	华凯建筑设计（上海）有限公司
内装设计单位	北京清尚建筑研究院有限公司
景观设计单位	深圳市铁汉生态环境股份有限公司
导向标识设计单位	北京广育德视觉技术有限公司
夜景照明设计单位	深圳市千百辉照明工程有限公司
弱电智能化设计单位	华体集团有限公司
外幕墙深化设计单位	深圳海外装饰工程有限公司

26 SANMENXIA WANDA PLAZA 三门峡万达广场

大商业施工图设计单位	北京东方国兴建筑设计有限公司
外立面设计单位	北京华雍汉维建筑咨询有限公司
内装设计单位	深圳市三九装饰工程有限公司
景观设计单位	北京中建设计院有限公司上海分公司
导向标识设计单位	北京视域四维城市导向系统规划设计有限公司
夜景照明设计单位	深圳市标美照明设计有限公司
弱电智能化设计单位	河南丹枫科技有限公司
外幕墙深化设计单位	北京嘉寓门窗幕墙股份有限公司

27 YICHUN WANDA PLAZA 宜春万达广场

大商业施工图设计单位	中国中轻国际工程有限公司
外立面设计单位	亚瑞建筑设计有限公司
内装设计单位	北京清尚建筑研究院有限公司
景观设计单位	上海帕莱登建筑景观咨询有限公司
导向标识设计单位	大连依斯特图文导视设计工程有限公司
夜景照明设计单位	深圳市名家汇科技有限公司
弱电智能化设计单位	北京国安电气有限责任公司
外幕墙深化设计单位	北京市金星卓宏幕墙工程有限公司

PART H PROJECT INDEX 项目索引

28 SUINING WANDA PLAZA 遂宁万达广场

大商业施工图设计单位	重庆市设计院
外立面设计单位	中外建工程设计与顾问有限公司
内装设计单位	中国中建设计集团有限公司
景观设计单位	重庆浩丰规划设计集团股份有限公司
导向标识设计单位	北京视域四维城市导向系统规划设计有限公司
夜景照明设计单位	深圳市金达照明有限公司
弱电智能化设计单位	上海中电电子系统科技股份有限公司
外幕墙深化设计单位	深圳蓝波绿建集团股份有限公司

29 XUZHOU TONGSHAN WANDA PLAZA 徐州铜山万达广场

大商业施工图设计单位	北京东方国兴建筑设计院有限公司
外立面设计单位	上海力夫建筑设计有限公司
内装设计单位	中国中建设计集团有限公司
景观设计单位	中外园林建设有限公司
导向标识设计单位	北京视域四维城市导向系统规划设计有限公司
夜景照明设计单位	北京清华同衡规划设计研究院有限公司
弱电智能化设计单位	江苏达海智能系统股份有限公司
外幕墙深化设计单位	中国建筑科学研究院有限公司

30 FOSHAN SANSHUI WANDA PLAZA 佛山三水万达广场

大商业施工图设计单位	广东省建筑设计研究院
外立面设计单位	北京华雍汉维建筑咨询有限公司
内装设计单位	深圳市三九装饰工程有限公司
景观设计单位	北京俪和景观工程设计有限公司
导向标识设计单位	大连依斯特图文导视设计工程有限公司
夜景照明设计单位	北京鱼禾光环境有限公司
弱电智能化设计单位	南京熊猫信息产业有限公司
外幕墙深化设计单位	金星卓宏幕墙工程有限公司

31 ZHENGZHOU HUIJI WANDA PLAZA 郑州惠济万达广场

大商业施工图设计单位	北京维拓时代建筑设计股份有限公司
外立面设计单位	深圳艾奕康上海分公司
内装设计单位	北京市建筑装饰设计院有限公司
景观设计单位	北京俪和景观工程设计有限公司
导向标识设计单位	深圳市上行线设计有限公司
夜景照明设计单位	艾德联合照明设计（北京）有限公司
弱电智能化设计单位	上海中电电子系统科技股份有限公司
外幕墙深化设计单位	中国建筑科学研究院有限公司

32 LIANXUNGANG WANDA PLAZA 连云港万达广场

大商业施工图设计单位	上海新外建工程设计与顾问有限公司
外立面设计单位	亚瑞建筑设计有限公司
内装设计单位	深圳市三九装饰工程有限公司
景观设计单位	华汇工程设计集团股份有限公司
导向标识设计单位	深圳市上行线设计有限公司
夜景照明设计单位	深圳市千百辉照明照明工程有限公司
弱电智能化设计单位	上海中电电子系统工程有限公司
外幕墙深化设计单位	北京和平幕墙工程有限公司

33 YINGKOU BAYUQUAN WANDA PLAZA 营口鲅鱼圈万达广场

大商业施工图设计单位	大连市建筑设计研究院有限公司
外立面设计单位	亚瑞建筑设计有限公司
内装设计单位	上海浦东建筑设计研究院有限公司
景观设计单位	上海帕莱登建筑景观咨询有限公司
导向标识设计单位	北京视域四维城市导向系统规划设计有限公司
夜景照明设计单位	北京清华同衡规划设计研究院有限公司
弱电智能化设计单位	大连理工科技有限公司
外幕墙深化设计单位	深圳市新山幕墙技术咨询有限公司

34 SHANGRAO WANDA PLAZA 上饶万达广场

大商业施工图设计单位	北京市建筑设计研究院有限公司
外立面设计单位	WSM Design limited 卫思蒙设计有限公司
内装设计单位	北京城邦建筑装饰工程有限公司
景观设计单位	中国城市建设研究院有限公司
导向标识设计单位	北京广育德视觉技术有限公司
夜景照明设计单位	福建福大建筑设计有限公司
弱电智能化设计单位	北京亚州卫星通信技术有限公司
外幕墙深化设计单位	厦门开联装饰工程有限公司

35 SHAOXING SHANGYU WANDA PLAZA 绍兴上虞万达广场

大商业施工图设计单位	华汇工程设计集团股份有限公司
外立面设计单位	熙林（北京）建筑设计咨询有限公司
内装设计单位	广东省美美设计工程有限公司
景观设计单位	中景汇设计集团有限公司
导向标识设计单位	宁波厚博交通设施科技有限公司
夜景照明设计单位	深圳市千百辉照明工程有限公司
弱电智能化设计单位	江苏达海智能系统股份有限公司
外幕墙深化设计单位	北京国科天创建筑设计院有限责任公司

36 CHAOYANG WANDA PLAZA 朝阳万达广场

大商业施工图设计单位	青岛北洋建筑设计有限公司
外立面设计单位	北京华雍汉维建筑咨询有限公司
内装设计单位	北京清尚建筑设计研究院有限公司
景观设计单位	中景汇设计集团有限公司
导向标识设计单位	北京视域四维城市导向系统规划设计有限公司
夜景照明设计单位	北京三色石环境艺术设计院有限公司
弱电智能化设计单位	北京国安电气有限责任公司
外幕墙深化设计单位	北京和平幕墙工程有限公司

37 SUZHOU WANDA PLAZA 宿州万达广场

大商业施工图设计单位	安徽省建筑设计研究院有限责任公司
外立面设计单位	上海帕莱登建筑景观咨询有限公司
内装设计单位	深圳市三九装饰工程有限公司
景观设计单位	华汇工程设计集团股份有限公司
导向标识设计单位	北京广育德视觉技术股份有限公司
夜景照明设计单位	深圳市金达照明有限公司
弱电智能化设计单位	华体集团有限公司
外幕墙深化设计单位	中国建筑科学研究院有限公司

38 CHENGDU QINGYANG WANDA PLAZA 成都青羊万达广场

大商业施工图设计单位	成都基准方中建筑设计有限公司
外立面设计单位	亨派建筑咨询（上海）有限公司
内装设计单位	北京市建筑装饰设计院有限公司
景观设计单位	上海帕莱登建筑景观咨询有限公司
导向标识设计单位	北京视域四维城市导向系统规划设计有限公司
夜景照明设计单位	上海易照景观设计有限公司
弱电智能化设计单位	北京国安电气有限责任公司
外幕墙深化设计单位	厦门开联装饰工程有限公司

39 DONGGUAN HUMEN WANDA PLAZA 东莞虎门万达广场

大商业施工图设计单位	奥意建筑工程设计有限公司
外立面设计单位	华凯建筑设计（上海）有限公司
内装设计单位	上海浦东建筑设计研究院有限公司
景观设计单位	深圳市铁汉生态环境股份有限公司
导向标识设计单位	北京视域四维城市导向系统规划设计有限公司
夜景照明设计单位	上海译格照明设计有限公司
弱电智能化设计单位	华体集团有限公司
外幕墙深化设计单位	深圳蓝波绿建集团股份有限公司

40 CHENGDU SHUANGLIU WANDA PLAZA 成都双流万达广场

大商业施工图设计单位	中国建筑西南设计研究院有限公司
外立面设计单位	艾奕康设计与咨询（深圳）有限公司
内装设计单位	北京市建筑装饰设计院有限公司
景观设计单位	深圳市致道景观设计有限公司
导向标识设计单位	大连依斯特图文导视设计工程有限公司
夜景照明设计单位	北京三色石环境艺术设计院有限公司
弱电智能化设计单位	上海智信世创智能系统集成有限公司
外幕墙深化设计单位	北京国科天创建筑设计院有限责任公司

41 BEIJING FENGTAI HIGH-TECH PARK WANDA PLAZA 北京丰台科技园万达广场

大商业施工图设计单位	北京柯德普建筑工程设计顾问有限公司
外立面设计单位	中国建筑设计院有限公司
内装设计单位	北京市建筑装饰设计院有限公司
景观设计单位	华汇工程设计集团股份有限公司
导向标识设计单位	北京四维城市导向系统规划设计有限公司
夜景照明设计单位	艾德联合照明设计（北京）有限公司
弱电智能化设计单位	大连理工科技有限公司
外幕墙深化设计单位	北京金星卓宏幕墙工程有限公司

42 BEIJING HUAIFANG WANDA PLAZA 北京槐房万达广场

大商业施工图设计单位	北京维拓时代建筑设计有限公司
外立面设计单位	中国建筑设计院有限公司
内装设计单位	北京市建筑装饰设计院有限公司
景观设计单位	华汇工程设计集团股份有限公司
导向标识设计单位	大连依斯特图文导视设计工程有限公司
夜景照明设计单位	栋梁国际照明设计（北京）中心有限公司
弱电智能化设计单位	北京益泰牡丹电子工程有限公司
外幕墙深化设计单位	北京国科天创建筑设计院有限责任公司

43 YANTAI DEVELOPMENT ZONE WANDA PLAZA
烟台开发区万达广场

大商业施工图设计单位	中国中轻国际工程有限公司
外立面设计单位	北京东方国兴建筑设计有限公司
内装设计单位	广东省集美设计工程公司
景观设计单位	鲁迅美术学院艺术工程总公司
导向标识设计单位	北京艺同博雅企业形象设计有限公司
夜景照明设计单位	艾德联合照明设计（北京）有限公司
弱电智能化设计单位	北京熊猫信息产业有限公司
外幕墙深化设计单位	北京和平幕墙工程有限公司

44 HAIKOU WANDA PLAZA
海口万达广场

大商业施工图设计单位	深圳市建筑设计研究总院有限公司
外立面设计单位	华凯建筑设计（上海）有限公司
内装设计单位	上海浦东建筑设计研究院有限公司
景观设计单位	鲁迅美术学院艺术工程总公司
导向标识设计单位	大连依斯特图文导视设计工程有限公司
夜景照明设计单位	深圳市金达照明有限公司
弱电智能化设计单位	南京熊猫信息产业有限公司
外幕墙深化设计单位	厦门嘉福幕墙铝窗有限公司

45 HUZHOU WANDA PLAZA
湖州万达广场

大商业施工图设计单位	华汇工程设计集团股份有限公司
外立面设计单位	中国新兴建设开发总公司
内装设计单位	中国中建设计集团有限公司
景观设计单位	中景汇景观设计有限公司
导向标识设计单位	华汇工程设计集团股份有限公司
夜景照明设计单位	深圳市标美照明设计工程有限公司
弱电智能化设计单位	河南丹枫科技有限公司
外幕墙深化设计单位	苏州柯利达装饰股份有限公司

46 SHANDONG BINZHOU WANDA PLAZA
山东滨州万达广场

大商业施工图设计单位	青岛北洋建筑设计有限公司
外立面设计单位	亚瑞建筑设计有限公司
内装设计单位	深圳市三九装饰工程有限公司
景观设计单位	深圳市致道景观设计有限公司
导向标识设计单位	北京视域四维城市导向系统规划设计有限公司
夜景照明设计单位	北京清华同衡规划设计研究院有限公司
弱电智能化设计单位	北京国安电气有限责任公司
外幕墙深化设计单位	中国新兴建设开发总公司

47 FUJIAN SANMING WANDA PLAZA
福建三明万达广场

大商业施工图设计单位	厦门合道工程设计集团有限公司
外立面设计单位	亚瑞建筑设计有限公司
内装设计单位	中国中建设计集团有限公司
景观设计单位	上海帕莱登建筑景观咨询有限公司
导向标识设计单位	北京视域四维城市导向系统规划设计有限公司
夜景照明设计单位	深圳市金达照明有限公司
弱电智能化设计单位	上海中电电子系统工程有限公司
外幕墙深化设计单位	厦门嘉福幕墙铝窗有限公司

48 LIUAN WANDA PLAZA
六安万达广场

大商业施工图设计单位	北京东方国兴建筑设计研究院有限公司
外立面设计单位	北京东方国兴建筑设计研究院有限公司
内装设计单位	北京清尚建筑设计研究院有限公司
景观设计单位	福建阿特贝尔景观设计有限公司
导向标识设计单位	北京视域四维城市导向系统规划设计有限公司
夜景照明设计单位	深圳市金达照明有限公司
弱电智能化设计单位	上海智信世创智能系统集成有限公司
外幕墙深化设计单位	北京国科天创建筑设计院有限责任公司

49 GUANGXI LIUZHOU LIUNAN WANDA PLAZA
广西柳州柳南万达广场

大商业施工图设计单位	广州宝贤华瀚建筑工程设计有限公司
外立面设计单位	艾奕康建筑设计（深圳）有限公司
内装设计单位	北京市建筑装饰设计院有限公司
景观设计单位	鲁迅美术学院艺术工程总公司
导向标识设计单位	北京视域四维城市导向系统规划设计有限公司
夜景照明设计单位	北京三色石环境艺术设计有限公司
弱电智能化设计单位	北京国安电气有限责任公司
外幕墙深化设计单位	厦门嘉福幕墙铝窗有限公司

INDEX OF WANDA HOTELS
万达独立酒店索引

01 WANDA REIGN SHANGHAI
上海万达瑞华酒店

大商业施工图设计单位	华东建筑设计研究院
外立面设计单位	福斯特事务所
内装设计单位	万达酒店设计研究院
景观设计单位	贝加艾奇（上海）建筑设计咨询有限公司 B+H
导向标识设计单位	CCDI 悉地国际
夜景照明设计单位	碧谱照明设计（上海）有限公司
弱电智能化设计单位	万达酒店设计研究院
外幕墙深化设计单位	上海凯腾幕墙设计咨询公司

02 WANDA VISTA ZHENGZHOU
郑州万达文华酒店

大商业施工图设计单位	中国核电工程有限公司
外立面设计单位	上海豪张思建筑设计有限公司
内装设计单位	万达酒店设计研究院
景观设计单位	中国建筑设计研究院
导向标识设计单位	北京四维城市导向系统规划设计有限公司
夜景照明设计单位	深圳市普莱思照明设计顾问有限责任公司
弱电智能化设计单位	万达酒店设计研究院
外幕墙深化设计单位	深圳市方大装饰工程有限公司

03 WANDA VISTA URUMQI
乌鲁木齐万达文华酒店

大商业施工图设计单位	中信建筑设计研究总院有限公司
外立面设计单位	EUmake 欧创建筑联合事务所有限公司
内装设计单位	万达酒店设计研究院
景观设计单位	深圳文科园林股份有限公司
导向标识设计单位	北京四维城市导向系统规划设计有限公司
夜景照明设计单位	北京鱼禾光环境设计有限公司
弱电智能化设计单位	万达酒店设计研究院
外幕墙深化设计单位	北京和平幕墙工程有限公司

04 WANDA REALM SIPING
四平万达嘉华酒店

大商业施工图设计单位	大连市建筑设计研究院
外立面设计单位	EUmake 欧创建筑联合事务所有限公司
内装设计单位	万达酒店设计研究院
景观设计单位	深圳文科园林股份有限公司
导向标识设计单位	成都奔流标识制作有限责任公司
夜景照明设计单位	北京洛西特灯光设计顾问有限公司
弱电智能化设计单位	万达酒店设计研究院
外幕墙深化设计单位	深圳市新山幕墙技术咨询有限公司

05 WANDA REALM XINING
西宁万达嘉华酒店

大商业施工图设计单位	甘肃省建筑设计研究院
外立面设计单位	华雍汉维设计院
内装设计单位	万达酒店设计研究院
景观设计单位	上海兴田建筑工程设计有限公司
导向标识设计单位	北京广育德视觉技术有限公司
夜景照明设计单位	北京三色石环境艺术设计有限公司
弱电智能化设计单位	万达酒店设计研究院
外幕墙深化设计单位	北京市金星卓宏幕墙工程有限公司

06 WANDA REALM BOZHOU
亳州万达嘉华

大商业施工图设计单位	安徽省建筑设计研究院有限公司
外立面设计单位	上海鼎实建筑设计有限公司
内装设计单位	万达酒店设计研究院
景观设计单位	北京中建建设设计院上海分院
导向标识设计单位	北京艺同博雅企业形象设计有限公司
夜景照明设计单位	央美光成(北京)建筑设计有限公司
弱电智能化设计单位	万达酒店设计研究院
外幕墙深化设计单位	中国新兴建设开发总公司

07 WANDA REALM YIWU
义乌万达嘉华酒店

大商业施工图设计单位	中国中轻国际工程有限公司
外立面设计单位	华凯建筑设计(上海)有限公司
内装设计单位	万达酒店设计研究院
景观设计单位	华东建筑设计研究院有限公司
导向标识设计单位	北京艺同博雅企业形象设计有限公司
夜景照明设计单位	深圳市标美照明设计工程有限公司
弱电智能化设计单位	万达酒店设计研究院
外幕墙深化设计单位	深圳市新山幕墙技术咨询有限公司

08 WANDA REALM SHANGRAO
上饶万达嘉华酒店

大商业施工图设计单位	北京市建筑设计研究院有限公司
外立面设计单位	上海思纳建筑规划设计有限公司
内装设计单位	万达酒店设计研究院
景观设计单位	中国城市建设研究院
导向标识设计单位	上海东楚装饰有限公司
夜景照明设计单位	福建福大建筑设计有限公司
弱电智能化设计单位	万达酒店设计研究院
外幕墙深化设计单位	厦门开联装饰工程有限公司

INDEX OF PROPERTIES FOR SALE
万达销售类物业索引

01 EXHIBITION CENTER OF WANDA CITY CHONGQING
重庆万达城展示中心

建筑设计单位	重庆市设计院
外幕墙设计单位	北京和平幕墙工程有限公司
内装设计单位	青岛腾远设计事务所有限公司
景观设计单位	广州山水比德景观设计有限公司
夜景照明设计单位	中国建筑装饰集团有限公司

02 EXHIBITION CENTER OF WANDA NO.1 CHENGDU
成都万达一号展示中心

建筑设计单位	浙江广厦建筑设计研究有限公司
外幕墙设计单位	厦门开联装饰工程有限公司
内装设计单位	上海腾申建筑规划设计有限公司
景观设计单位	中国建筑设计院有限公司
夜景照明设计单位	北京鱼禾光环境设计有限公司

03 HEPING EXHIBITION CENTER OF WANDA CITY GUILIN
桂林万达城和平展示中心

外建筑设计单位	青岛腾远设计事务所有限公司
外幕墙深化设计单位	青岛腾远设计事务所有限公司
内装设计单位	上海青荷建筑装饰有限公司
景观设计单位	北京北林地景园林规划设计院有限责任公司
夜景照明设计单位	青岛腾远设计事务所有限公司

04 DEMONSTRATION AREA OF YONGCHUAN WANDA PALACE
永川万达华府示范区

建筑设计单位	九源建筑设计有限公司
外幕墙设计单位	北京国科天创建筑设计院有限责任公司
内装设计单位	华诚博远(北京)建筑规划设计有限公司
景观设计单位	中国建筑设计院有限公司
夜景照明设计单位	深圳市千百辉照明工程有限公司

05 PROTOTYPE SHOP DEMONSTRATION AREA OF WANDA CITY CHENGDU - GREEN TOWN & WATER STREET
成都万达城商铺样板示范区——青城水街

建筑设计单位	成都基准方中建筑设计有限公司
外幕墙设计单位	四川华西建筑装饰工程有限公司
景观设计单位	广州怡境景观工程有限公司
夜景照明设计单位	四川大光明城市照明工程有限公司

06 COMMERCE & RESIDENCE BLOCK OF ZHANGZHOU TAISHANG WANDA PLAZA
漳州台商万达广场商墅街区

建筑设计单位	厦门佰地建筑设计有限公司
外幕墙设计单位	厦门佰地建筑设计有限公司
景观设计单位	厦门华旸建筑工程有限公司
夜景照明设计单位	普莱思照明设计顾问有限公司

万达商业规划

2016
WANDA COMMERCIAL PLANNING

叶宇峰 朱其玮 冯腾飞 方伟 侯卫华 风雪昆 沈文忠 李浩 季元
马红 范珑 陈海亮 曹春 阎红伟 黄引达 罗沁 王玉龙 刘江
兰勇 刘佩 陆峰 宋锦华 张振宇 高振江 李斌 曹国峰 章宇峰
葛宁 黄勇 方芳 蒲峰 赵陨 黄路 苏仲洋 朱欢 张德志 屈娜
孙佳宁 赵青扬 郭扬 徐立军 李小强 王宇石 李彬 张宁 王朝忠
张堃 董明海 石路也 赵剑利 秦鹏华 路清淇 范群立 王文广
黄川东 赵洪斌 蓝毅 王翔 张顺 周明 吴凡 吕鲲 李万勇
张黎明 宋雷 张晓冬 黄涛 高建航 袁喆 陆轩 桑伟 贺明 王清文
罗贤君 冯晓芳 王少雷 孙穆元 闵盛勇 林涛 高霞 宋永成 谭瑶
熊厚 朱迪 余斌 孟晗 晁志鹏 陈杰 李民伟 姚建刚 虞朋 王吉
王云 Marcus 杨琳 王进纯 钟文渊 Quentin 张涛 朱广宁 张佳
周澄 Maria 都晖 陈勇 王奕 张争 孙海龙 刘晓波 Sofia 丰佳
韦云 李易 邓金珂 杨春龙 闫颇 党恩 陈玭潭 何志勇 刘向阳 殷超
卫立新 刘锋 钟光辉 赵海滨 方文奇 王凯 邹洪 李华 刘俊
张鳃 路滨 王静 罗冲 张鹤 庞博 张琳 白宝伟 杨汉国 王永磊
陈理力 李捷 杨娜 段堃 胡存珊 徐小莉 齐宗新 杨艳坤
罗琼 孟祥宾 王睿麟 张洋 栾海 陈志强 任志忠 刘易昆
吴晓璐 梅海斌 康宇 马雪健 于春雨 李韦达 叶萍 栗冬

尹强 门瑞冰 张东光 杨旭 毛晓虎 王群华 王福魁 孙培宇 曾静
陈文娜 石亮 刘大伟 昌燕 马刚 王治天 黄建好 张鹏翔 马长宁
刘安 张悦 董华维 刘江 薛勇 叶啸 武春雨 胡延峰 万志斌
薛瑜 刘征 文善平 王权 荣万斗 周映晗 周志超 俞小华 周升森
安云泽 栾赫 覃韬 邵强 张克 潘亮 李暄荣 林彬 洪剑 李子强
郭晨光 杨磊 赵宁宁 韩茂俊 李涛 彭亚飞 孙一琳 漆国强 张雪晖
李春阳 周鹏 李万顺 王凡 梁国涛 魏大强 庞庆 李江涛 刘晓敏
顾东方 宋波 原红波 陈涛 霍雪影 冯董 马申申 车心达 程欢
潘鸿岭 韦胜平 戚士林 张蕾 袁文卿 纪文青 龚芳

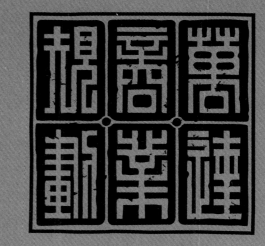

图书在版编目（CIP）数据

万达商业规划 2016：汉英对照 / 万达商业规划研究院，万达商业地产设计中心，万达商业地产技术研发部 主编 . – 北京：中国建筑工业出版社，2018.1
ISBN 978-7-112-21748-9

Ⅰ.①万… Ⅱ.①万…②万…③万… Ⅲ.①商业区－城市规划－中国－汉、英 Ⅳ.①TU984.13

中国版本图书馆 CIP 数据核字 (2018) 第 003184 号

责任编辑：徐晓飞　张　明　封　毅　张瀛天
执行编辑：刘易昆
美术编辑：陈　唯
英文翻译：喻蓉霞　王晓卉　郝　婧
责任校对：焦　乐

万达商业规划 2016

万达商业规划研究院
万达商业地产设计中心　　主编
万达商业地产技术研发部
*
中国建筑工业出版社出版、发行（北京海淀三里河路 9 号）
各地新华书店、建筑书店经销
北京雅昌艺术印刷有限公司制版印刷
*
开本：787×1092 毫米　1/8　印张：38$\frac{1}{2}$　字数：612 千字
2018 年 1 月第一版　　2018 年 1 月第一次印刷
定价：900.00 元
ISBN 978-7-112-21748-9
　　　（31594）
版权所有　翻印必究
如有印装质量问题，可寄本社退换
（邮政编码　100037）